Introductory Guide to Design-of-Experiments (DoE)

Wilbert O.L. Sibanda (PhD)

BSc Life Sciences (Wits, South Africa), **BSc (Med) Hons Pharmacology** (UCT, South Africa), **MSc (Med) Pharmacy** (Wits, South Africa), **PhD Information Technology** (NWU, South Africa)

Biostatistician, School of Nursing and Public Health, College of Health Sciences, University of KwaZulu-Natal, Durban, South Africa, 4001

(Introductory Guide to Design-of-Experiments (DoE)
Copyright © 2017 by (Wilbert O.L. Sibanda)

ISBN 978-0-620-80830-9

9   780620   808309

All rights reserved. No part of this book may be reproduced or transmitted in any form or by any means without written permission from the author.

Printed in South Africa by Wilbert Sibanda

ISBN **978-0-620-80830-9**

Dedication

This book is lovingly dedicated to:

My wonderful wife Cathrine Sibanda, my children Lorraine, Janice, Njabulo, Vuyo Sibusisiwe and my grand-daughter Khanyisile Sibanda. Special honor to my late parents Mr. and Mrs. Sibanda.

From dad Dr. Wilbert 'Banquo' Sibanda, January, 2016

DoE Vocabulary

Factor: This is one of the independent variables under investigation in the experiment

Level: These the different values the factor can take in an experiment

Experimental Run: This is the process of conducting an experiment with factors set at certain values

Response: This is the numerical outcome of an experimental run

Full Factorial Experiment: This is an experimental design used to determine the effect of all possible combinations across all levels of the factor under the study

Fractional Factorial Experiment: Experiment designed to study a given number of factors with the fraction of the experimental runs required for the full factorial experiment

Blocking: This is the removal from experimental error a contributor of variability identified but not under investigation

Table of Contents

Foreword

I hope this book will serve as a great introduction to this wonderful world of Design-of-Experiments (DoE) and I hope it will inspire you to start using Design-of-Experiments (DoE) in your next project so you can unlock the extraordinary power of Design-of-Experiments (DoE). *Good luck*

Preface

The purpose of this handbook is to introduce Design-of-Experiments (DoE) including how to make experiments efficiently, describe how to analyze the data, how to interpret results and suggest how to optimize experiments. Reading this book will give you a practical understanding of Design-of-Experiments (DoE)

A Short History of Design-of-Experiments

Ronald A. Fisher is considered the founding father of Design-of-Experiments (DoE), between the years 1920 and 1930, while working at Rothamsted Experimental Station, an agricultural research station in London (*Yates, 1964*).

Sir Ronald Fisher

(**Source:** *https://www.famousscientists.org/ronald-fisher/*)

Ronald fisher demonstrated how reliable results could be obtained from an experiment in the presence of extraneous variables. Even though, DoE was initially used in agriculture, it has recently found application in industry to improve quality (*Ilzarbe, 2008*), engineering (*Taguchi, 1987*), military (*Yildirim, (2008)* aircraft design (*Giunta, (1997)*, and pharmaceutical industry (*Sibanda, 2004*).

George Box is credited with the development of the experimental procedure for optimizing chemical processes using Response Surface Methodology (RSM) *(Box, 1987)*. An in-depth description of the design of experiments theory can be found in Montgomery *(Montgomery, 1995.)* and Montgomery (1997).

Genichi Taguchi, a Japanese is renowned for his application of Design-of-Experiments to attain robust designs that reduced variation and improved product quality *(Taguchi, 1986)*. Robust design select factor settings that are associated with low variability.

In recent years, DOE is increasingly being recognized as a vital tool to improve process yield, reduce variability, development time and overall costs.

References

Yates, F. (1964). Sir Ronald Fisher and the design of experiments. Biometrics, 20(2), 307-321).

Ilzarbe, L., Álvarez, M. J., Viles, E., & Tanco, M. (2008). Practical applications of design of experiments in the field of engineering: a bibliographical review. Quality and Reliability Engineering International, 24(4), 417-428

Taguchi, G., & Taguchi, G. (1987). System of experimental design; engineering methods to optimize quality and minimize costs (No. 04; QA279, T3.)

Yildirim, U. Z., Sabuncuoglu, I., Tansel, B., & Balcioglu, A. (2008, December). A design of experiments approach to military deployment planning problem. In Simulation Conference, 2008. WSC 2008. Winter (pp. 1234-1241). IEEE).

Giunta, A. A. (1997). Aircraft multidisciplinary design optimization using design of experiments theory and response surface modeling methods (p. 185). Blacksburg, VA: Virginia polytechnic institute and state university)

Sibanda, W., Pillay, V., Danckwerts, M. P., Viljoen, A. M., van Vuuren, S., & Khan, R. A. (2004). Experimental design for the formulation and optimization of novel cross-linked oilispheres developed for in vitro site-specific release of Mentha piperita oil. AAPS PharmSciTech, 5(1), 128-141.

Box, G. E., & Draper, N. R. (1987). Least squares for response surface work. Response Surfaces, Mixtures, and Ridge Analyses, Second Edition, 29-91.

Montgomery, D. C., & Myers, R. H. Response Surface Methodology: Process and Product Optimization Using Designed Experiments, 1995.

Taguchi, G. (1986). Introduction to quality engineering: designing quality into products and processes.

Chapter 1: Introduction to Design-of-Experiments (DoE)

This chapter introduces the concept of Design-of-Experiments (DoE), outlines the uses of DOE as well as advantages of DoE compared to traditional (One-Factor-at-a-Time) OFAT methods of development of experiments. I will also discuss the various data analysis steps in DoE. Towards the end of this chapter, I will also explore the process of problem formulation in DoE.

1.1. Introduction to Design-of-Experiments (DoE)

Design-of-Experiments (DoE) is fundamentally a methodical process of constructing, administering and analysis of an experiment to identify factors that influence the experimental process. It is therefore, important to design an experiment in an appropriate manner, to fully appreciate the different independent processes affecting the response variable. DoE is sequence of experiments where input variables of a process and their effects on response variables are measured. DoE is a powerful tool for maximising knowledge and information obtained from an experimental process, while minimising the amount of resources used in the experiment.

1.2. Important Principles in DoE

The three main principles of DoE are:

a) *Replication*-Replication allows the investigator to determine experimental error

b) *Randomization*-This is the random process of assigning individuals to different treatment groups. The purpose of randomization is to ensure that there is no bias, through ensuring that effects of confounding variables, whether known or unknown are evenly divided across treatment groups.

c) *Blocking*-This is the process of arranging sampling units in similar groups in terms of one or more covariates. Treatments are then given in a random manner in those groups. This ensures that variability within blocks is less than variability between blocks. Therefore, blocking is conducted to increase precision by removing the effect of known nuisance factors, such as batch to batch variability.

Therefore, blocking and randomization both seek to minimize bias from confounding factors.

Example of Blocking

A study was conducted to determine the efficacy of a new drug developed to treat rheumatoid arthritis. Patients were assigned into blocks based on age as shown in Table below.

Age (years)	Placebo	Rheumatoid Treatment
30-40	100	100
41-50	100	100
51-60	100	100

In each age-group (block), patients were randomly assigned to treatment. In the above trial, 100 patients in each age-group received placebo and another 100 in each age-group received the new rheumatoid treatment. Therefore, in this design, each treatment condition has equal numbers of patients in each age-group. This design removes age as a potential source of variability and as a potential confounding variable.

1.3. Five Stages of DoE

a) Conceptualization stage

This stage of the DoE process involves the formulation of the research idea and the identification of the purpose of the experiment. In other words, the researcher interrogates the purpose of the research. During this stage, the researcher identifies the factors to be investigated. A well-thought out experiment results in a better understanding of the research process. By extension, a properly planned experiment allows the researcher to conduct experiment accurately and with much ease. A badly planned experiment results in unreliable data.

b) Screening

Screening designs are used to identify the important factors from a large number of factors. Main factors identified using a screening design can be used for further analysis. Therefore, screening designs are primarily focused on main effects, not interactions of factors.

c) Response Surface Design

Response surface methodologies (RSM) are used to determine the optimal response within specified ranges of RSM. RSM identify optimal settings for the main or significant factors identified in the screening design. The objective of an RSM is to either increase or decrease process variability and identify optimal conditions that achieve both. Therefore, an RSM model has the capability to determine curvature, identify interactions among factors and optimize the process. The quadratic model used to estimate response surface is the quadratic model:

$$Y = \beta_0 + \sum_{i=1}^{P} \beta_i x_i + \sum_{i=1}^{P}\sum_{j=1}^{P} \beta_{ij} x_i x_j + \beta_{11} x^2_1 + \sum_{i=1}^{p} \beta_{ii} x^2_i + \varepsilon \tag{i}$$

Where

β_0 = overall mean square

$\beta_i = main\ effect\ of\ each\ factor\ (i = 1, 2, ...p)$ β_{ij} = two-way interaction between the ith and jth factors and

β_{ii} = the quadratic effect for the ith factor.

d) Robustness Testing

Robustness refers to the degree to which a process or system operates correctly in the presence of exceptional inputs or stressful environmental conditions. Robustness testing is a methodology for detecting the vulnerabilities of a component (Lei, 2010).

The uncontrolled inputs are sometimes referred to as noise. A thorough understanding of the origins of noise ensures that development of process or product takes into account the robustness of the process to these extraneous factors.

1.4. Conducting and Analyzing Experiments

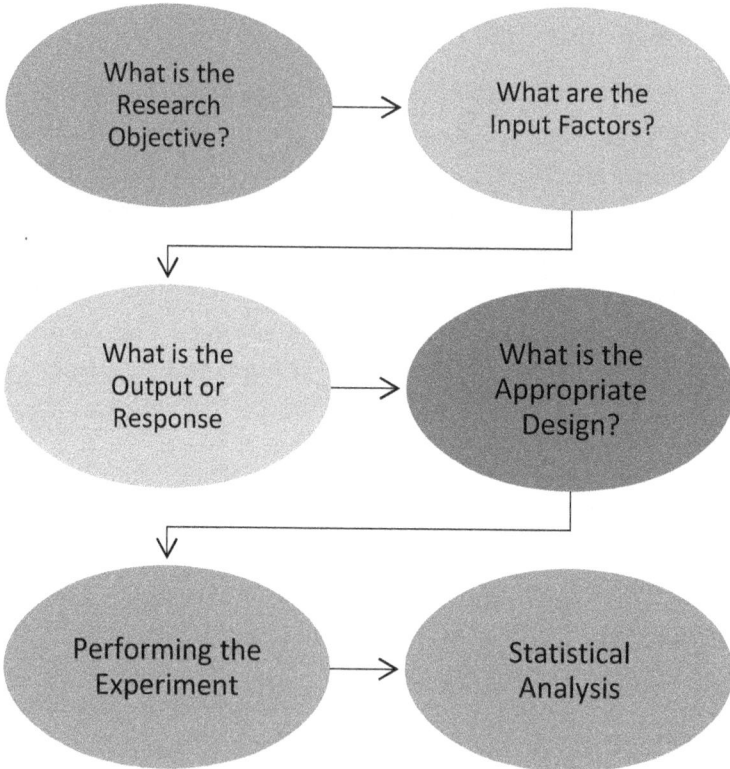

Fig. 1.1: Design-of-experiments flow-chart

a) What is the Research Objective?

This is the crucial stage in DoE. It deals with the formulation of the research problem and the creation of objectives of the study. Therefore, a clearly defined research problem enables the investigator to rapidly assess whether the study is relevant.

b) What are the Input Factors and their levels?

These are the factors that will be controlled during the experiment, including their appropriate levels, i.e. 2 level, 3-level etc.

c) What is the Output or response?

This is the effect variable, frequently referred to as the dependent variable. Response variables measure the effect of the independent variables on the experimental process.

d) What is the appropriate design?

There are different types of design-of-experiments (DoE). Some examples of experimental designs include, One-Factor-At-A-Time (OFAT) design, Factorial designs and Response surface designs.

(i)   One-Factor-At-A-Time (OFAT)

Traditionally experiments were designed to study the effect of one factor on one response. This was called the one-factor-at-a-time (OFAT) approach as it depicted an experimental approach that sought to investigate one factor at a time as opposed to investigating all the factors at the same time.

Disadvantages of the OFAT system

- OFAT requires more experimental runs compared to DoE for the same precision in response prediction,
- OFAT is unable to estimate factor interactions,

- OFAT is unable to identify the optimal settings of the independent variables. Therefore varying a single factor at a time does not result in a real optimum, resulting in different outcomes for different starting points.

Advantages of DoE over OFAT Experiments

- DoE requires less resources in the form of experiments, time and materials compared to OFAT,
- DoE tends to provide more precise estimates of the effects of independent factors on the response,
- DoE enables the systematical and logical estimation of interactions between factors. OFAT is unable to estimate factor interactions.
- DoE explores a larger region of the factor space compared to OFAT and thus enabling the efficiency of prediction of the response in the factor space through the reduction of the variability of the estimation of the response in the factor space. The latter situation enables more efficient process optimization.

(ii)　Factorial Designs

A single study with numerous independent variables is called a factorial design. This means that a factorial design involves simultaneously more than one factor each time at two or more levels. By definition, factorial designs have several factors simultaneously affecting the same characteristic under study. Therefore, varying the levels of the factors at the same time, is efficient in terms of time and cost of the experiment, while at the same time allowing for the investigation of interactions between factors. Examples of factorial designs are, full factorial designs, fractional designs and Plackett-Burman designs, proposed by R.L. Plackett and J.P. Burman in 1946 (Plackett, R. L., & Burman, J. P. (1946).

(iii)　Response Surface Designs

These are the collection of mathematical and statistical techniques for empirical model building. These techniques seek to optimize a response that is affected by several independent variables. These models are used to produce a prediction model to determine curvature and interactions among factors and to optimize the process. In order, to fit a quadratic model, at least three levels for each input variables are needed, i.e. high, medium and low levels (Telford, (2007). Examples of these designs are the central composite and Box-Behnken designs.

e) Performing the Experiment

This is the process of collecting scientific data on the response to an intervention in a systematic way to maximize the chance of answering a question correctly. Therefore, experiments should be planned before they start and planning should include the statistical methods used to assess the results.

f) Statistical Analysis

Appropriate techniques should be used to analyze the data and the analysis should indicate the aim of the study. The purpose of an exploratory analysis should be to identify patterns in the data.

1.5.   Overview of Data Analysis

Case-Study 1-A DoE Example

A study was conducted to develop and optimize an in-vitro site-specific delivery system for the treatment of irritable bowel syndrome (IBS), using a Plackett-Burman design. A novel cross-linked calcium-aluminium-alginate-pectinate oilisphere complex was used as a potential system for the in-vitro site-specific release of Mentha piperita, an essential oil used for the treatment of irritable bowel syndrome (IBS). The physicochemical and textural properties of the crosslinked complexes such as unhydrated matrix resilience, matrix hardness and total fracture energy were measured (Sibanda et. al, 2004).

(a) *State the independent variables*

In this case-study, the independent variables were:

Independent variables	Type	Levels
Sodium alginate	Polysaccharide	0-1.5% wt/vol
Pectin	Polysaccharide	0-1.5% wt/vol
Calcium chloride	Salt	0-4% wt/vol
Aluminium chloride	Salt	0-4% wt/vol
Cross-linking reaction time	Time	0.5-6 hours

(b) *State the dependent variables*

In this study, the response variables were:

Response variable	Units
Unhydrated matrix resilience	%
Matrix hardness	N/m
Total fracture energy	J

(c) *Plackett-Burman experimental design*

The design matrix used in this case-study was the Plackett-Burman, a special category of two-level designs where not all factor level combinations are considered. The purpose of the design matrix is to demonstrate the experimental plan. The design matrix contains independent variables in a regression model which aims to explain the observed data

(d) Regression model

Formulation	Independent Variables						Response Variables		
	SA	P	CaCl$_2$	AlCl$_3$	Al(SO4)$_3$	CRT	UR	TFE	MH
1	0	1.5	4	4	0	6	8.48	0.01	674.55
2	1.5	0	4	0	0	0.5	3.08	0.014	-329.75
3	1.5	0	0	0	4	6	7.97	0.012	6142.50
4	0	1.5	0	0	0	6	0	0.009	-1504.70
5	1.5	1.5	0	4	0	0.5	102.01	0.014	30530.60
6	1.5	1.5	0	4	4	0.5	71.02	0.016	37846.40
7	0	0	0	0	0	0.5	0	0.009	-4859.60
8	1.5	0	4	4	0	6	47.64	0.009	-343.20
9	0	0	4	4	4	0.5	0	0.003	3286.40
10	0.75	0.75	2	2	2	3.25	53.72	0.011	7461
11	0.75	0.75	2	2	2	3.25	53.78	0.011	7461
12	0	0	0	4	4	6	0	0.002	2442.80
13	1.5	1.5	4	0	4	6	94.95	0.018	10341
14	0	1.5	4	0	4	0.5	19.65	0.011	6654.7

N.B. SA=Sodium alginate, P=Pectin, CaCl$_2$, AlCl$_3$=Aluminium chloride, Al$_2$(SO4)$_3$ = Aluminium sulphate, CRT=Cross-linking Reaction Time, UR=Unhydrated Matrix Resilience, TFE=Total Fracture Energy, MH=Matrix Hardness

Regression equations were developed to describe the statistical relationship between independent variables and the dependent variables in Case-study 1 for the three responses:

Unhydrated Matrix Resilience (%)

$$= 0.38 + 0.012 * [SA] + 0.031 * [P] + 0.94 * [CaCl_4] + 0.28 * [AlCl_3] + 0.73 * [Al_2(SO_4)_3]$$

$$+ 0.69 * CRT \text{ (i)}$$

Matrix Hardness (N/m)

$$= 0.62 + 0.053 * [SA] + 0.058 * [P] + 0.17 * [CaCl_4] + 0.14 * [AlCl_4] + 0.24 * [Al_2(SO_4)_3]$$

$$+ 0.14 * CRT \text{ (ii)}$$

Total fracture Energy (J)

$$= 0.80 + 0.0006 * [SA] + 0.011 * [P] + 0.08 * [CaCl_4] + 0.60 * [AlCl_4] + 0.37 * [Al_2(SO_4)_3]$$

$$+ 0.62 * CRT \text{ (iii)}$$

e) Interpretation of the regression model

P-values and regression coefficients are some of the tools used to analyze regression equations. The p-value for each regression term can be used to test the null hypothesis that the coefficient is equal to zero. In other words, the p-value tests if a coefficient has no significant effect. Therefore, a low p-value below 0.05 demonstrates that the null hypothesis can be rejected, implying that the independent variable has a significant contribution to the response variable. On the other hand, regression coefficients represent the mean change in the dependent variable for every one unit change in the independent variable, while keeping all the other independent variables constant.

1.5.   Formulation of a DoE problem

1.5.1. Definition of the problem

The first step in the formulation of a problem should involve the identification of the research problem and the explanation of its importance in the study.

1.5.2. Definition of experimental procedure
a. Screening Design

The purpose of the screening design is to identify the most influential factors and their appropriate ranges on the experimental response. Screening designs are mainly used to identify the important factors that affect the process out of a large number of other factors. The factors identified in the screening design can be used to conduct further experimental analysis to understand the manufacturing process.

b. *Optimization Design*

The purpose of the optimization design is to develop the best experimental settings for the main factors that were identified in the screening objective in order to achieve the desired outcome. The optimization objective is therefore aimed at either increasing the response output or reducing the process variability or simultaneously achieve both.

c. *Robustness testing*

The optimization objective set out to obtain the best experimental conditions for the production of a final product. Following the identification of the best experimental conditions, it is becomes important to ensure that the product or process is insensitive to variations during product manufacture (Micskei, 2012). Robustness is the degree to which a system operates correctly in the presence of stressful environmental conditions. The variations could result from changes to factors beyond the control of the investigator that affect the manufacturing process. Examples of such extraneous factors in a manufacturing process could be humidity, ambient temperature and variations in materials. It is essential for the researcher to fully understand the possible sources of product or process variation, in order be able to develop techniques to ensure the process is insensitive to these extraneous factors, frequently called noise.

1.5.3. *Definition of Independent variables*

This involves the categorization of experimental factors into the following categories:

a) *Controllable and uncontrollable variables*

Controlled variables are factors whose settings can be controlled by the investigator, such as ordinary process factors, such as pH, temperature and pressure. Uncontrollable variables are factors whose settings cannot be controlled.

b) *Qualitative and quantitative variables*

In general, quantitative factors are continuous factors that can be measured numerally. In simple language, quantitative factors are measurable quantities. Qualitative factors are categorical measurements, whose values are not numerical. Examples of categorical measurements are sex of an individual (male or female) or disease status (infected or uninfected).

1.5.4. *Description of response variables*

It is important that the experimenter identifies the appropriate response(s) for the experimental process. Responses can be continuous, discrete or semi-continuous.

1.5.5. *Identification of the DoE type*

a) *Screening Designs*
These DoE methodologies are suitable for linear and interaction models. Examples are fractional factorial and Plackett-Burman designs.

b) *Optimization designs*
These DoE methodologies are suitable for quadratic models. Examples of quadratic models are Central composite and Box-Behnken designs.

1.5.6. *Selection of a regression model*

There are fundamentally three main types of polynomial regression models

a) Linear regression model

Linear models describe a continuous independent variable as a function of one or more dependent variables. These models enable the understanding of how a quantitative dependent variable depends on one or more independent variables. However, the independent variables may be quantitative or qualitative. The general equation for a linear model is:

$$Y = \beta 0 + \beta_{1x_1} + \beta_2 x_2 + \dots + \beta_{p-1} x_{p-1} + \varepsilon \qquad \text{(Eqn. iv)}$$

Where:

Y is a linear function of independent variables, $x_1 \dots x_{p-1}$.

β and ε represent linear parameter estimates and error terms respectively. Examples of linear regression models include:

i.   Simple linear regression: These models use only one independent variable,

ii.   Multiple linear regression: These models are characterized by multiple independent variables

iii. Multivariate linear regression: Multivariate linear regression models are comprised of multiple dependent or response variables.

b) Interaction regression model

The term interaction refers to how the effect on the response of one independent variable depends on the level of one or more other independent variables. Therefore, the interaction of factors arises when the effect of one independent variable depends on the value of another explanatory variable.

Example of Interaction of Factors

In CASE-STUDY 1, where a Placket-Burman design was used to develop an in-vitro site-specific drug delivery system for the treatment of irritable bowel syndrome (IBS). An interaction plot is used to determine the effect size of interactions using the data in the Plackett-Burman matrix. The results of the study are shown below.

Plackett-Burman Design matrix

Formulation	Independent Variables						Response Variables		
	SA	P	CaCl4	AlCl3	AlSO4	CRT	UMR	TFE	MH
1	0	1.5	4	4	0	6	8.48	0.01	674.55
2	1.5	0	4	0	0	0.5	3.08	0.014	-329.75
3	1.5	0	0	0	4	6	7.97	0.012	6142.50
4	0	1.5	0	0	0	6	0	0.009	-1504.70
5	1.5	1.5	0	4	0	0.5	102.01	0.014	30530.60
6	1.5	1.5	0	4	4	0.5	71.02	0.016	37846.40
7	0	0	0	0	0	0.5	0	0.009	-4859.60
8	1.5	0	4	4	0	6	47.64	0.009	-343.20
9	0	0	4	4	4	0.5	0	0.003	3286.40
10	0.75	0.75	2	2	2	3.25	53.72	0.011	7461.00
11	0.75	0.75	2	2	2	3.25	53.78	0.011	7461.00
12	0	0	0	4	4	6	0	0.002	2442.80
13	1.5	1.5	4	0	4	6	94.95	0.018	10341.00
14	0	1.5	4	0	4	0.5	19.65	0.011	6654.7

Where: SA = Sodium alginate, P = Pectin, CaCl4 = Calcium Chloride, AlCl3 = Aluminium Chloride, CRT = Crosslinking reaction time, UMR = Unhydrated Matrix Resilience, TFE = Total Fracture Energy, MH = Matrix Hardness

Interaction Table

Interaction	Average Response Variables		
	UMR	TFE	MH
SA⁻P⁻	0	0.003	289.9
SA⁺P⁻	19.6	0.012	1823.2
SA⁻P⁺	9.38	0.010	1941.50
SA⁺P⁺	89.33	0.016	26239.33

Where: SA⁻ = Low level Sodium Alginate, SA⁺=High level Sodium Alginate, P⁻=Low level Pectin, P⁺=High level Pectin

Exp #	SA	P	CaCl4	AlCl3	Al2SO4	CRT	UMR
1	0	1.5	4	4	0	6	8.48
2	1.5	0	4	0	0	0.5	8.08
3	1.5	0	0	0	4	6	7.97
4	0	1.5	0	0	0	6	0
5	1.5	1.5	0	4	0	0.5	102.01
6	1.5	1.5	0	4	4	0.5	71.02
7	0	0	0	0	0	0.5	0
8	1.5	0	4	4	0	6	47.64
9	0	0	4	4	4	0.5	0
10	0.75	0.75	2	2	2	3.25	53.72
11	0.75	0.75	2	2	2	3.25	53.72
12	0	0	0	4	4	6	0
13	1.5	1.5	4	0	4	6	94.95
14	0	1.5	4	0	4	0.5	19.65

Calculate Average of factors SA and P at different levels

	SA-P Avg
SA⁻P⁻	0
SA⁺P⁻	19.6
SA⁻P⁺	9.38
SA⁺P⁺	89.33

Interaction Plot

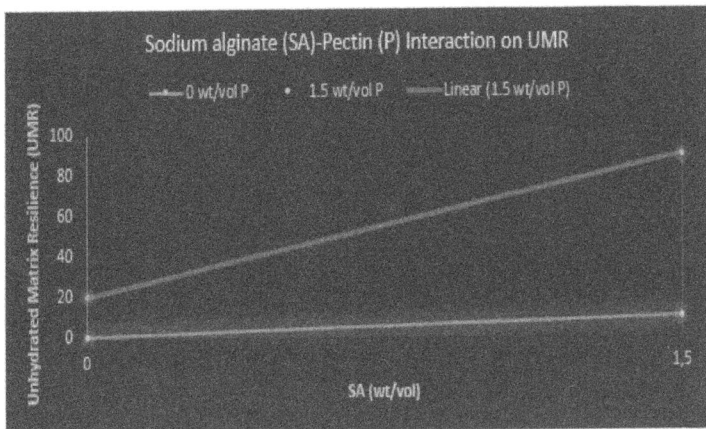

Sodium alginate (SA)-Pectin (P) Interaction on UMR

Legend: 0 wt/vol P · 1.5 wt/vol P — Linear (1.5 wt/vol P)

Unhydrated Matrix Resilience (UMR)

SA (wt/vol)

Key: SA⁻=Sodium alginate low level, SA⁺=Sodium alginate high level, P⁻=Pectin low level, P⁺=Pectin high level

Conclusion:

If the data for responses for any two-factors are plotted and the results show non-parallel lines, this means there is interaction between the two factors. However, a statistical test is required to determine if the interaction is significant. Therefore, in the above example, there is an interaction between Sodium alginate and Pectin, based on the convergence of the two lines!

b) Response: Total Fracture Energy (TFE)

Exp #	SA	P	CaCl4	AlCl3	Al2SO4	CRT	TFE
1	0	1.5	4	4	0	6	0.01
2	1.5	0	4	0	0	0.5	0.014
3	1.5	0	0	0	4	6	0.012
4	0	1.5	0	0	0	6	0.009
5	1.5	1.5	0	4	0	0.5	0.014
6	1.5	1.5	0	4	4	0.5	0.016
7	0	0	0	0	0	0.5	0.003
8	1.5	0	4	4	0	6	0.009
9	0	0	4	4	4	0.5	0.003
10	0.75	0.75	2	2	2	3.25	0.01
11	0.75	0.75	2	2	2	3.25	0.01
12	0	0	0	4	4	6	0.002
13	1.5	1.5	4	0	4	6	0.018
14	0	1.5	4	0	4	0.5	0.01

	SA-P Avg
SA⁻P⁻	0.003
SA⁺P⁻	0.012
SA⁻P⁺	0.010
SA⁺P⁺	0.016

Interaction Plot

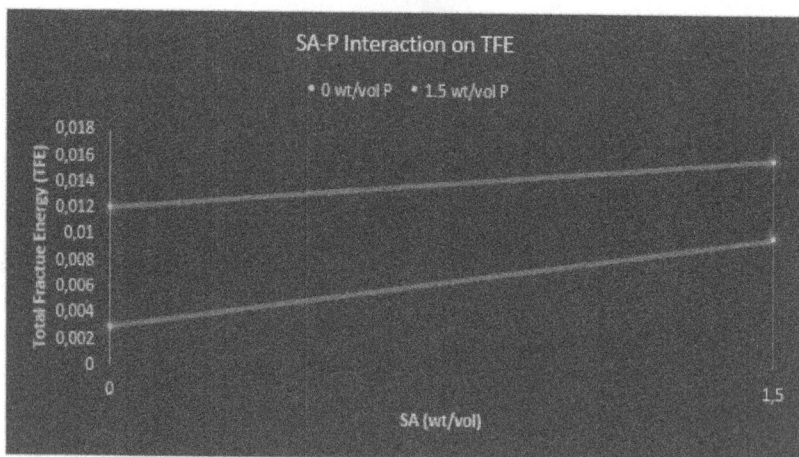

SA-P Interaction on TFE

• 0 wt/vol P   • 1.5 wt/vol P

Key: SA⁻=Sodium alginate low level, SA⁺=Sodium alginate high level, P⁻=Pectin low level, P⁺=Pectin high level

Conclusion

The plots are converging, suggesting the possible existence of interaction between SA and P to influence Total Fracture Energy.

N.B. Therefore, an interaction between factors occurs whenever two factors, acting together, produce mean differences that are not explained by the main effects of the two factors. When an interaction is large, the corresponding main effects have little practical meaning. This means that a significant interaction tends to mask the significance of main effects.

26

c) *Quadratic Regression model*

A quadratic model is a polynomial function in one or more variables where the highest term is of the second degree. A univariate single-variable quadratic model has the form:

$$f(x) = ax^2 + bx + c,$$

where a, b, and c are constants

References

1. Sibanda, Wilbert, et al. "Experimental design for the formulation and optimization of novel cross-linked oilispheres developed for in vitro site-specific release of Mentha piperita oil." AAPS PharmSciTech 5.1 (2004): 128-141.

2. Sibanda, Wilbert, and Philip Pretorius. "Comparative study of the application of central composite face-centred (CCF) and Box–Behnken designs (BBD) to study the effect of demographic characteristics on HIV risk in South Africa."Network Modeling Analysis in Health Informatics and Bioinformatics 2.3 (2013): 137-146.

3. Lei, B., Li, X., Liu, Z., Morisset, C., & Stolz, V (2010). Robustness testing for software components. Science of Computer Programming, 75(10), 879-897.

4. Plackett, R. L., & Burman, J. P. (1946). The design of optimum multifactorial experiments. Biometrika, 33(4), 305-325.

5. Telford, J. K. (2007). A brief introduction to design of experiments. Johns Hopkins apl technical digest, 27(3), 224-232.

6. Micskei, Z, Madeira, H., Avritzer, A., Majzik, I., Viera, M. & Antunes, N (2012). Robustness testing techniques and tools. In Resilience Assessment and Evaluation of Computing Systems (pp. 323-339). SpringerBerlin, Heidelberg, 2012.

Inspirational Invention

James West-Father of the Foil Electret Microphone

Born in Prince Edward County, Virginia, on February 10, 1931, James West attended Temple University before working for Bell Labs. Along with Gerhard M. Sessler, he developed the foil electret microphone, an inexpensive, compact device that is now used in 90 percent of all contemporary microphones.

Chapter 2: Full Factorial Design

This chapter introduces the basic principles of full factorial designs and outlines the advantages of full factorial designs. This chapter will also define main effects of factors. The end of this chapter, will be dedicated to the analysis of data using least squares as well as explanation of the meaning of coefficients of regression models.

2.1. Basic principles of full factorial

Full factorial design is an experimental design with all possible combinations of levels from two or more factors. Factorial designs can be used with any number of factors and factor levels. However, a very high number of factors creates a problem in the interpretation of three-way interactions. Full factorial designs are building blocks of many design-of-experiment models. Full factorial designs can be used to build more complex composite designs for response surface models such as central composite designs. Full factorial designs are also used to develop fractional factorial designs. Full factorial designs attempt to conduct all possible combinations of all levels of all factors. The total number of experiments for a full factorial are:

$$Total\ Number\ of\ Experimental\ Runs = m^k$$

Where $m = number\ of\ levels\ and\ k = number\ of\ factors$

2.2. Important facts about factorial design

a) Effect Hierarchy Principle

This states that lower order effects are more likely to be important than higher order effects. By extension, effects of the same order are equally likely to be important.

b) Effect Sparsity Principle (Box-Meyer)

This is similar to the Pareto effect and stipulates that the number of relatively important effects in a factorial experiment is small. Effect hierarchy and Sparsity Principles find more relevance in screening *designs!*

c) Effect Heredity Principle (Hamada-Wu)

This principle states that in order for an interaction to be significant at least one of its parent factors should be significant.

2.3. Advantages of two-level full factorial designs

Full factorial designs enable the determination of the effect of each factor and their interactions. Two-level full factorial designs are important for screening research, due to its ability to estimate interaction effects of factors. An important feature of two-level factorial designs is that they are balanced and orthogonal.

A balanced design is an experimental design where all experimental and factor level combinations have the same number of observations. An orthogonal design is an experimental design with factors that are independent of each other, and at right angles to each other i.e. Factor A*Factor B= (1)(1)+(1)(-1)+(-1)(1)+(-1)(-1)=0 (see Table 2.1). This means that Factor A is estimated independently from Factor B.

In general full factorial designs are ideal for 2-4 factors, while experiments with 5 or more factors are suited for fractional factorial designs.

Table 2.1. Example of Two-level Full factorial Design (2^2)

Factor A	Factor B
+1 (High)	+1 (High)
+1 (High	-1 (Low)
-1 (Low)	+1 (High)
-1 (Low)	-1 (Low)

2.4. Disadvantages of two-level full factorial

The cost of the experiment is elevated due to the number of factors or levels.

Examples of two-level and three-level Full Factorial Designs

Table: Two-Level Full factorial Design

Experiment	Factor A (Level)	Factor B (Level)
1	+1 (High)	-1 (Low)
2	+1 (High)	+1 (High)
3	-1 (Low)	-1 (Low)
4	-1 (Low)	+1 (High)

Table: Three-Level Full Factorial

Experiment	Factor A	Factor B	Factor C
1	+1 (High)	-1 (low)	-1 (Low)
2	+1 (High)	-1 (Low)	+1 (High)
3	+1 (High)	+1 (High)	+1 (High)
4	+1 (High)	+1 (High)	-1 (Low)
5	-1 (Low)	-1 (Low)	-1 (Low)
6	-1 (Low)	-1 (Low)	+1 (High)
7	+1 (High)	+1 (High)	+1 (High)
8	+1 (High)	+1 (High)	-1 (Low)

2.5. Definition of main effect of a factor

A main effect of a factor is the effect of an independent variable on the dependent variable, while ignoring the effects of all the other predictor variables. Therefore, main effect is the main difference between the levels of one factor. It is therefore a plot of the means of the response for each level of a factor, allowing for the determination of which main effects are important.

Case-Study 2 (Main Effects)

A study was conducted to determine drug release from polymeric crosslinked drug delivery systems. Two drug delivery systems consisting of PVA polymer were studied for drug release at two different pH settings, namely pH 1.5 and 6.8 over three time periods, 20, 40 and 60 mins. The results are shown in the Table below:

Time (mins)	Fractional Drug Release	
	pH 1.5	pH 6.8
20	0.40	0.55
40	0.60	0.80
60	0.75	0.95

Calculation of Main Effect

The main effect of an independent variable at each pH, is the effect of the variable averaging over all levels of other variables in the experiment. The main effects of PVA pH1.5 and PVA pH 6.8 are assessed by computing the mean for the two levels of PVA averaging fractional release across all three time periods.

The mean for PVA pH 1.5 is: $(0.40 + 0.60 + 0.75) / 3 = 0.58$

The mean for PVA pH 6.8 is: $(0.55 + 0.80 + 0.95) / 3 = 0.77$

Therefore, the main effect of   a) PVA pH 1.5 is **0.58**

b) PVA pH 6.8 is **0.77**

Analysis-of-Variance (ANOVA) provides a significance test for the main effect of PVA pH 1.5 and PVA pH 6.8. If the main effect of type of PVA is significant, the null hypothesis that there is no difference between PVA pH 1.5 and PVA pH 6.8 would be rejected.

Case-Study 3: Calculation of main effects for a two level factorial design by HAND

A study was conducted to determine the effects of women's ages and their male sexual partner's ages on the risk of HIV infection. The ages of women were divided into two groups namely; ≤ 20 (coded -1) and 21-30 years (coded +1), while the ages of men were divided into ≤ 24 (coded -1) and 25-34 (coded +1), as shown in Table 2.2. For both men and women the lower age was coded (-1) and the higher age (+1).

Table 2.2. Example of Two-level Full factorial Design (2^2)

Demographic characteristic	Data coding	
	-1 (Lower level)	+1 (high level)
Woman's age (years)	≤ 20	≥ 30
Male sexual partner's age (years)	≤ 24	≥ 34

A two-level factorial design matrix was developed as shown in Table 2.3.

Table 2.3. Two-level full factorial Design (2^2)

Experiment	Demographic characteristics (Factors)		HIV risk (Response)
	Woman's age	Male sexual partner's age	
1	+1	+1	0.36
2	+1	-1	0.21
3	-1	+1	0.13
4	-1	-1	0.14

Data Source: *Sibanda et. al., 2012*

Fig. 2.1: Case-studies

Solution to Case-study 3

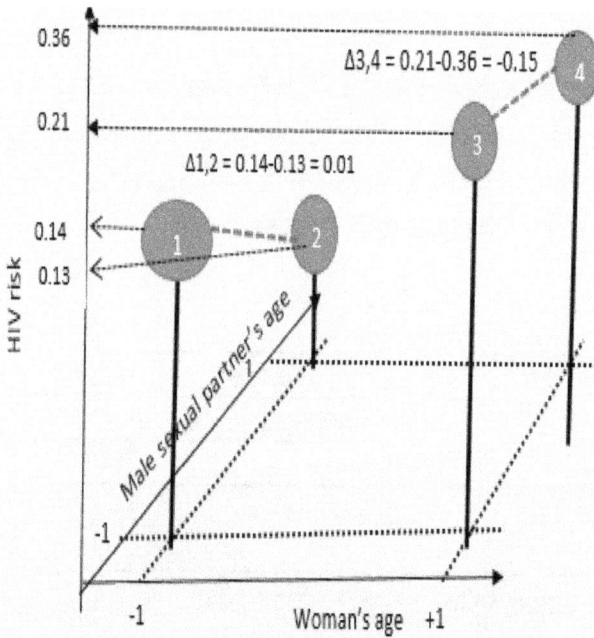

Main effect of male sexual partner's age = $(\Delta1,2 + \Delta3,4)/2 = -0.14$

Main effect of woman's age = $(\Delta1,3 + \Delta2,4)/2 = 0.15$

Fig. 2.2: Calculation of main effects of variables by HAND

2.6. Application of Linear Regression for Data Analysis

A regression model is suitable for the analysis of DoE data using least squares fit. The primary aim of the least squares model is to minimize the vertical distance between itself and the observed points on a scatterplot.

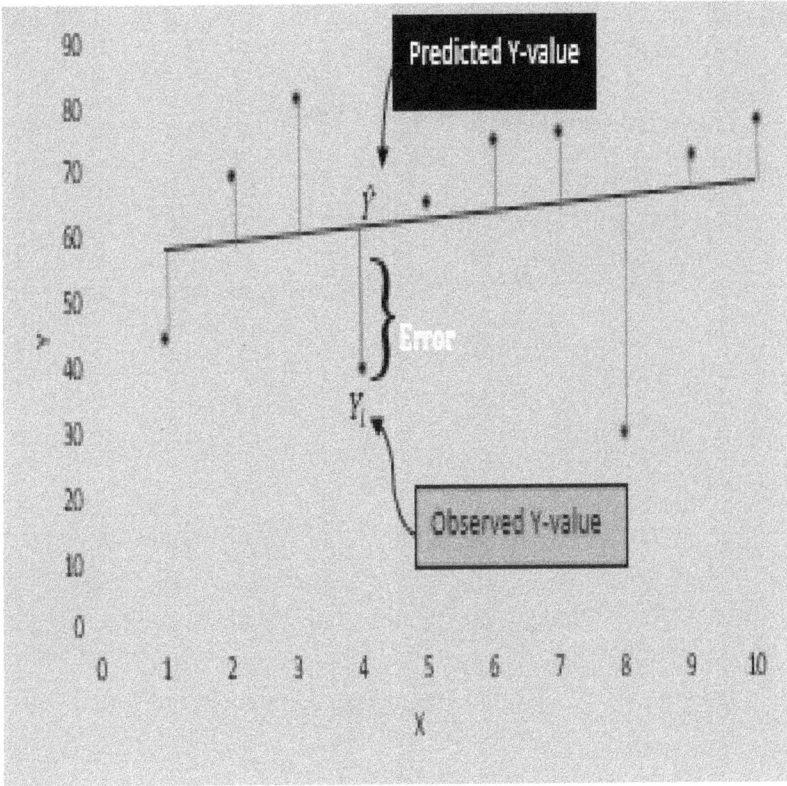

Fig. 2.3: Least Squares Model

The Y value predicted by the regression model is given by $\hat{Y} = b_0 + b_1 X_i$ and the error of prediction is calculate as follows:

$$\text{error} = Y_i - \hat{Y} = Y_i - (b_0 + b_1 X_i),$$

where (X_i, Y_i) represents the observed data

Therefore the best line that fits the data is represented by the expression that minimizes the sum of errors between observed and predicted response values $\sum_{1}^{n} (Y_i - \hat{Y}_i)$. By extension, the best regression line that fits the data is represented by the line that minimizes the sum of squared errors between

35

observed and predicted response values and this is called ordinary least squares regression (OLS).

Least Squares Line

$$Y = b_0 + b_1 x$$

Where x is independent variable or regressor,

Y is dependent variable and b_0 is intercept

Given $(x_1, y_1), (x_2, y_2), \ldots (x_n, y_n)$

Slope $\quad b_1 = r \dfrac{S_y}{S_x}$

Y intercept $\quad b_0 = \bar{y} - b\bar{x}$

Where $S_x = \sqrt{\dfrac{\displaystyle\sum_{i=1}^{n}(x_i - \bar{x})^2}{n-1}}$ = Standard deviation of $x_1,,x_n$

Where $S_y = \sqrt{\dfrac{\displaystyle\sum_{i=1}^{n}(y_i - \bar{y})^2}{n-1}}$ = Standard deviation of $y_1,,y_n$

$$r = \dfrac{\displaystyle\sum_{i=1}^{n}(x_i - \bar{x})((y_i - \bar{y})}{(n-1)S_x S_y}$$ = Correlation between x and y

$$SSE = \sum_{i=1}^{n} y^2_{\ i} - b_0 \sum_{i=1}^{n} y_i - b_1 \sum_{i=1}^{n} x_i y_i = Error\ Sum\ of\ Squares$$

Case-Study 4-Practical Application of Least Squares Fit

A study was conducted to investigate the changes in HIV prevalence rates amongst Generation X black women attending antenatal clinics across the nine provinces of the Republic of South Africa over a period of ten years (Sibanda, CIBB 2015).

Solution to Case-Study 4

Table 2.4

Year (X)	HIV rate/1 000 women (Y)	XY	X²	$\hat{Y}=292.29+14.88x$	$Y_i - \hat{Y}$
1	285.9	285.9	1	307.17	-21.27
2	328.57	657.14	4	322	6.57
3	320.87	962.61	9	336.93	-16.06
4	371.04	1484.16	16	351.81	19.23
5	393.50	1967.5	25	366.69	26.81
6	389.98	2339.88	36	381.57	8.41
7	394.07	2758.49	49	396.45	-2.38
8	403.37	3226.96	64	411.33	-7.96
9	417	3753	81	426.21	-9.21
10	436.95	4369.50	100	441.09	-4.14
$\sum Y = 55$	$\sum Y = 3741.3$	$\sum XY = 21805.1$	$\sum X^2 = 385$		$\sum (Y_i - \hat{Y}) = 0$

The equation of the least squares model is:

$$Y = a + bX$$

In order to calculate the values of a and b, we need to create simultaneous equations as follows;

$$\sum y = na + b \sum x \qquad (i)$$

$$\sum xy = a \sum x + b \sum x^2 \qquad (ii)$$

Using the data in Table 2.4 in equations (i) and (ii) gives equations (iii) and (iv) below:

$$3741.3 = 10a + 55b \qquad (iii)$$

$$21805.1 = 55a + 385b \qquad (iv)$$

Solving for a and b using substitution method gives equations v

$$a = 374.13 - 5.5b \qquad (v)$$

Substituting equation (v) in equation (iv) gives

$$21805.1 = 55(374.13 - 5.5b) + 385b \qquad (vi)$$

$$b = 14.88$$

To calculate a, the value of $b = 14.88$ is substituted in equation (v) giving

$$a = 374.13 - 5.5(14.88)$$

$$a = 292.29$$

Therefore the values of a and b are: $a = 292.29$ and $b = 14.88$

Therefore the least squares regression model is: $\hat{Y} = 292.29 + 14.88x \qquad (vii)$

2.7. Advantages of least squares regression

Least squares regression models:

- (i)   are robust to slight changes in factor levels,
- (ii)   are capable of estimating the residual error,
- (iii)   can be used in regions of the experimental space where experiments are not feasible and
- (iv)   are comprised of coefficients that indicate the influence of factors

2.8. The meaning of regression model coefficients

An example of a regression model is given in equation $(viii)$. The regression model has two predictor variables (x_1) and (x_2) and one response variable Y. ε Represents the residual error.

$$Y = b_0 + b_1 x_1 + b_2 x_2 + \varepsilon \quad (viii)$$

The parameters in the regression model $(viii)$ are; $b_0 = Y$ intercept, $b_1 =$ 1st regression coefficient, and $b_2 =$ 2nd regression coefficient

2.9. Interpretation of the intercept

The Y intercept (b_0) represents the value of the response variable when both predictors x_1 and x_2 are equal to zero.

A study using Central Composite-Face-Centered (CCF) and Box-Behnken Designs (BBD) to determine the main and interaction effects of demographic characteristics on the risk of HIV infection amongst pregnant women attending antenatal clinics in South Africa, the following expression was developed;

$Risk\ of\ HIV\ infection$

$$= 0.33 + 0.15 * woman's\ age$$

$$+ 0.002 * male\ partner's$$

$$- 0.09 * woman's\ education$$

39

$$+ 0.005 * Parity \text{ (ix)}$$

The Y intercept is 0.33 provided the woman's age, male partner's age, woman's educational level and parity are all equal to zero. This means that when all the predictor variables are zero, the risk of HIV infection is 0.33. However, it is not possible to have all the predictor variables equal to zero. Therefore the intercept is has no practical meaning in this instance. The intercept has a meaningful interpretation if it is reasonable that all the independent variables in the regression model can be equal to zero.

2.10. Interpretation of coefficients of continuous predictor variables

Using equation (ix) to explain the coefficients of continuous variables

Risk of $HIV\ infection\ =\ \ \ \ \ \ 0.33\ +0.15 * woman's\ age$

$\ \ \ \ \ \ \ \ \ + 0.002 * male\ partner's\ age$

$\ \ \ \ \ \ \ \ \ \ \ \ - 0.09 * woman's\ education\ leve$

$+ 0.005 * Parity$

All the predictor variables in equation (ix) are continuous variables, i.e. woman's age, male partner's age and woman's educational level. Taking the woman's age as an example, 0.15 represents the difference in the predicted risk of HIV infection for each one unit difference in the woman's age provided the male partner's age, woman's educational level and parity remain constant. In other words, this means that if the woman's age differed by unit, male partner's age, woman's educational level and parity did not differ, the risk of HIV infection of the woman will differ by 0.15 units on average.

2.11. Interpretation of coefficients of categorical variables

Case-Study 5-Example of Interpretation of Coefficients of Categorical data

A two-predictor logistic regression model was fitted to data to test the research hypothesis regarding the relationship between the likelihood that an inner city child is recommended for remedial reading instruction and his or her reading score and gender (Peng, 2002).

$$Predicted\ logit\ of\ remedial = 0.54 + (-0.026) * Reading + 0.648 * Gender$$

According to the model, the log of the odds of a child being recommended for remedial reading instruction was negatively related to reading scores ($p<0.05$) and positively related to gender ($p<0.05$).

In other words, the higher the reading score, the less likely it is that a child will be recommended for remedial reading classes than girls because boys were coded to be +1 and the girls 0. In fact, the odds of a boy being recommended for remedial reading programs were 1.91 times greater than odds for girls.

2.12. Interpretation of the interactions of a regression model

In Case-Study 3, a three-level Box-Behnken design was used to study the main and interaction effects of demographic characteristics on the risk of HIV infection amongst women attending antenatal clinics in South Africa (Sibanda, 2012).

Table 2.6: Coding the data

(Factors) Demographic characteristics of women in antenatal clinics	Levels		
	Low (-1)	Middle (0)	High (+1)
Woman's age (years)	≤20	21-29	≥30
Male partner's age (years)	≤24	25-33	≥34
Parity	0	1	≥2
Woman's educational level (Grades)	≤8	9-11	12-13

Using the coded variables in Table 2.6, a Box-Behnken design was used to develop a regression model for the prediction of risk of HIV infection based on a woman's demographic characteristics.

$Risk\ of\ HIV\ infection = 0.32 + 0.18 * woman's\ age - 0.02 * male\ partner's\ age +$

From the equation, the coefficient of the interaction term between woman's age and male partner's age is 0.13. Therefore, based on the interaction between woman's age and male partner's age, the effect of woman's age on risk of HIV infection is shown by equation (xii) below:

$Risk\ of\ HIV\ infection = 0.18 + 0.13 * male\ partner's\ age$ $\qquad (xii)$

a) Women with male partners younger than 24 years (coded -1), the risk of HIV infection is

$$0.18 + 0.13 * (-1) = 0.05 = 5\%\ (xiii)$$

b) Women with male partners older than 34 years (coded 1), the risk of HIV infection is

$$0.18 + 0.13 * (1) = 0.31 = 31\%\ (xiv)$$

Therefore, the interaction of factors ensures that the effect of the woman's age on the risk of HIV depends on the age of her male sexual partner.

Solved Problems

Problem 1: Select from the list below, a design-of-experiments in which all factors levels are combined with all factor levels of every other factor

a) Full factorial design
b) Fractional factorial design
c) Response surface design
d) Screening design

Solution

a) Full factorial design

The full factorial design investigates all the factors in an experiment. The full factorial looks at all possible combinations that are associated with the factors and their levels.

The full factorial design focusses on the main factors and their interactions on the measured response

In a full factorial experiment, all possible combinations of the factors are investigated at all levels tested. This means that an experiment with p factors and q levels, would have q^p experimental runs i.e.

An experiment with:

(i) 3 factors at 2 levels has $2^3 = $ **8 experimental runs**

(ii) 3 factors at 3 levels has $3^3 = $ **27 experimental runs**

(iii) 4 factors at 2 levels has $4^2 = 16 \; experimental \; runs$

Problem 2: A pharmaceutical company wishes to develop a cough mixture in its research and development division. The three input factors (excipients) are sulphur, cetrimide and methyl paraben as shown in Table 2.7. The purpose is to determine the importance of each of the excipients on the final product.

43

Table 2.7: High (+1) and Low (-1) settings for the excipients of the cough mixture

Excipients	Low (-1)	High (+1)	Units
Sulphur	7	14	g
Cetrimide	0.05	0.10	g
Methyl Paraben	0.1	0.2	m/m

Show the following:

a) Graphical representation of the factor settings
b) Using a full factorial 2 level design, indicate the number of experimental runs
c) Hypothetical full factorial model for three factors at two levels each
d) Demonstrate full factorial design matrix showing experimental runs and factor settings
e) What is the purpose of replication in a design-of-experiments matrix
f) What is homogeneity of variance
g) Comment on randomization

Solution

a) Graphical representation of factor settings

The purpose is to determine various combinations of factor settings in order to obtain the best cough mixture. There are eight different ways of combining high and low settings of the cough mixture excipients, Sulphur (x_1), cetrimide (x_2) and methyl paraben (x_3).

44

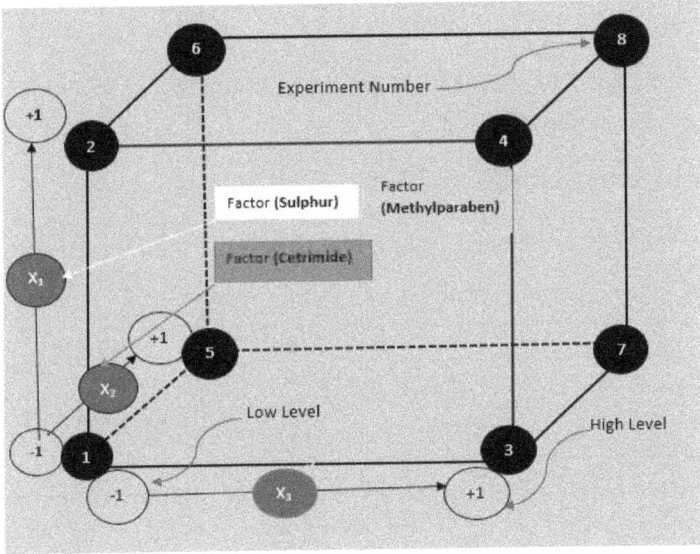

Fig. 2.4: Full factorial experiment with 3 factors at two levels each

b) Number of Experimental Runs

In general, for a full factorial experiment with k factors at two levels, there will be 2^k combinations of the different levels.

In this example, (k=3) and the total number of experimental runs is: $2^3 = 8$

c) If all the possible combinations of factors are run, this will result in the estimation of all the main and interaction effects as follows:
   - ✓ 3 x main effects (Sulphur (x_1), cetrimide (x_2) and methyl paraben (x_3).
   - ✓ 3 x two-factor interactions (x_1x_2, x_1x_3, x_2x_3)
   - ✓ 1 x three-factor interactions ($x_1x_2x_3$)

Therefore, a full factorial design model has all the eight β coefficients $\{\beta_0 \cdots \beta_{123}\}$;

$$y = \beta_0 + \beta_1 x_1 + \beta_2 x_2 + \beta_3 x_3 + \beta_{12} x_1 x_2 + \beta_{13} x_1 x_3 + \beta_{23} x_2 x_3 + \beta_{123} x_1 x_2 x_3 + \varepsilon \quad (xv)$$

d) Full factorial design matrix showing experimental runs and factor settings

The numbers of the vertices of the box in Fig. 2.4 indicate the standard order of the experiments as shown in Table 2.8.

Table 2.8: Full factorial design matrix and factor settings

Experimental order	x_1 (Sulphur)	x_2 (Cetrimide)	x_3 (Methyl paraben)
1	-1	-1	-1
2	+1	-1	-1
3	-1	+1	-1
4	+1	+1	-1
5	-1	-1	+1
6	+1	-1	+1
7	-1	+1	+1
8	+1	+1	+1

e) Purpose of Replication

Running the full factorial design more than once provides information on the variability and consistency of the response. It is important to control variability to ensure validity of the data. Randomization, blocking and replication are central to ensuring results are replicable. In general, noise factors that are not controlled can be handled by randomization (to avoid bias), replication (to increase precision) and blocking (to isolate noise).

f) Homogeneity of variance

Homogeneity of variance assumes that the dispersion of the response is uniform across the entire experimental space, i.e. $H_0: \sigma^2_1 = \sigma^2_2 = \ldots = \sigma^2_n$ (Games, 1972). The concept of assumption of homogeneity of variance assumes that all comparison groups in a study have equal variance. Some of the tests used for homogeneity of variance are Hartley's F_{max}, Cochran's, Levene's and Bartlett's tests. Levene's test is the usual F-test for Levene's equality of means, to test the null hypothesis that the variance is equal across groups (Carroll, 1985). A p-value less 0.05 indicates a violation of the assumption. Replication of the full factorial design allows the investigator to test the assumption of homogeneity of variance across the experimental space!

g) Randomization of experiments

Randomization of the experiment reduces the effect of external factors on the response. It is therefore advisable to randomize the experiments as many times as possible. However, randomization does not mean hapharzardness (Altman, 1991)

Problem 3:

A company conducted a study to determine the optimal combination of ingredients in the manufacture of an affordable laxative mixture for the relief of constipation. Preliminary studies indicated that the optimal formulation could be found between the following concentration ranges of pharmaceutical ingredients: sodium sulphate anhydrous (150-250mg), sodium sulphate $10H_2O$ (450-550mg), Chloroform (2-4 %v/v) and ethyl alcohol (3-5 %v/v). The laxative effect (response) was measured on a scale between 0 and 100.

Show:

a) Using a two-level full factorial design develop a design matrix
b) Analysis of experimental data using:
   (i)   Normal probability plot
   (ii)   Box plot
   (iii)   Histogram
c) Create the theoretical model
d) Fit the model to data
e) Test the model assumptions using residual graphs
f) Show the important main and interaction effects

Table 2.9: Factor settings for the laxative mixture

Factor	Factor Name	Factor Levels	Low	High

			Level	Level
1	Sodium sulphate anhydrous	2	150 (-1)	250 (+1)
2	Sodium sulphate hydrous	2	450 (-1)	550 (+1)
3	Chloroform	2	2 (-1)	4 (+1)
4	Ethyl alcohol	2	3 (-1)	5 (+1)

Solution:

a) Two-level full factorial design matrix

Table 2.10: A two level full factorial design matrix with four factors

Exp No	Run Order	Sodium sulphate anhy (x1)	Sodium sulphate hy (x2)	Chloroform (X3)	Ethyl alcohol (x4)	Lux effect (Response)
1	14	150	450	2	3	11
3	18	150	500	2	3	51
5	5	150	450	3	3	75
7	13	150	500	3	3	60
9	16	150	450	2	4	14
11	8	150	500	2	4	71
13	1	150	450	3	4	80
15	19	150	500	3	4	80
17	9	150	450	2	3	13
18	11	150	450	2	3	12
19	2	150	450	2	3	14
2	3	200	450	2	3	70
4	6	200	500	2	3	70

48

6	4	200	450	3	3	79
8	15	200	500	3	3	68
10	12	200	450	2	4	34
12	10	200	500	2	4	80
14	17	200	450	3	4	40
16	7	200	500	3	4	56

A two level full factorial design with four factors was used to develop a design matrix with $2^4 = 16$ experimental runs. Each experimental run was tested for its laxative effect as shown in Table 2.10.

b) Validity of error assumptions

In general an assumption is made that all error terms are identical and independently normally distributed with average value of zero.

b (i) Normal probability plot

The normal probability plot is used to test the normality of errors terms. When the data points are distributed around the diagonal line, it means that the assumption of normality of error terms is satisfied. Therefore, a normal probability plot is used to identify outliers, skewness, kurtosis and a need for transformation. There are different patterns of residual distribution, i.e. non-normality (shown by a non-straight line), skew (distribution curved at the tails) and outliers (points that lie far away from the lines). Residuals are defiend as the difference between observed and predicted values of the response variable, i.e. $Residual = y - \hat{y}$,

Where:

$$y = observed\ value\ of\ dependent\ variable$$

$$\hat{y} = predicted\ value\ of\ dependent\ variable$$

$$Standardised\ residual\ i = \frac{Residual_i}{Standard\ Deviation\ of\ Residual_i}$$

49

where standardised residual, has a **mean = 0** and **standard deviation =1**.

In this example, it can be concluded that the error terms are approximately normal.

a) (i) Normal Probability Plot

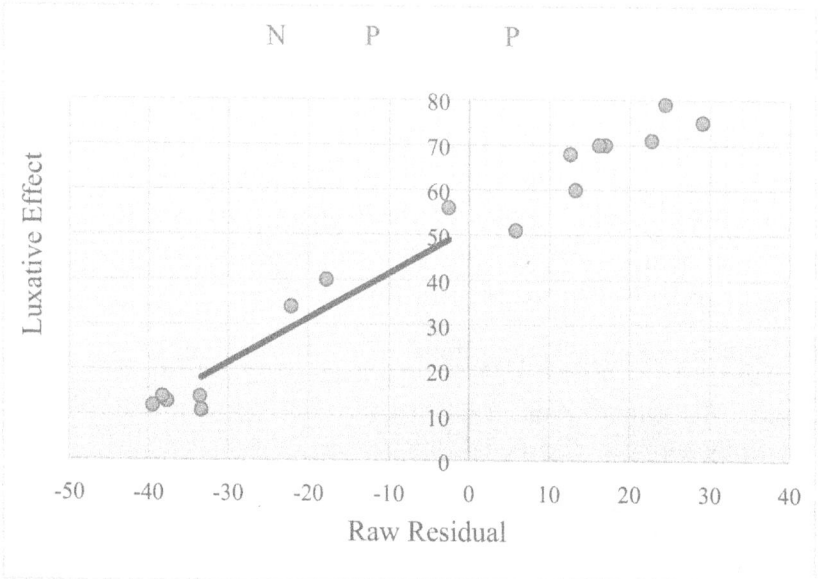

(a)

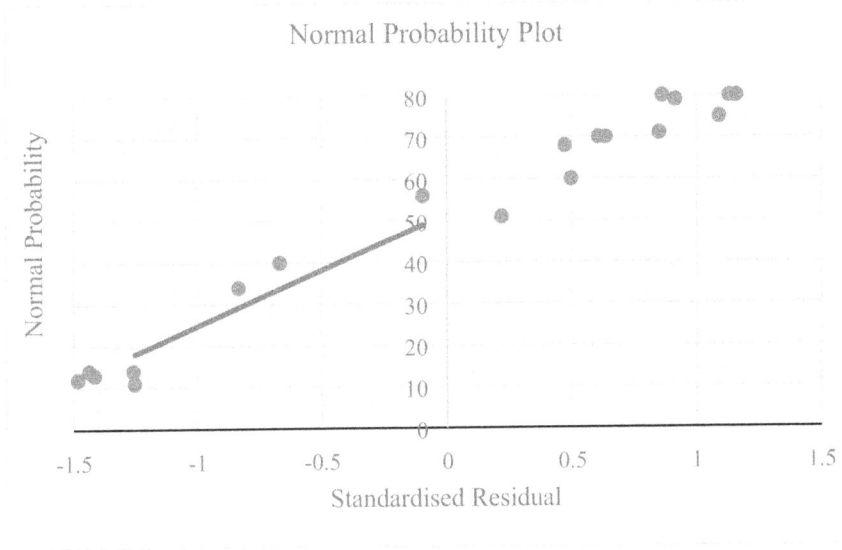

Fig. 2.5: Normal probability plot of a) residuals and b) standardised residuals

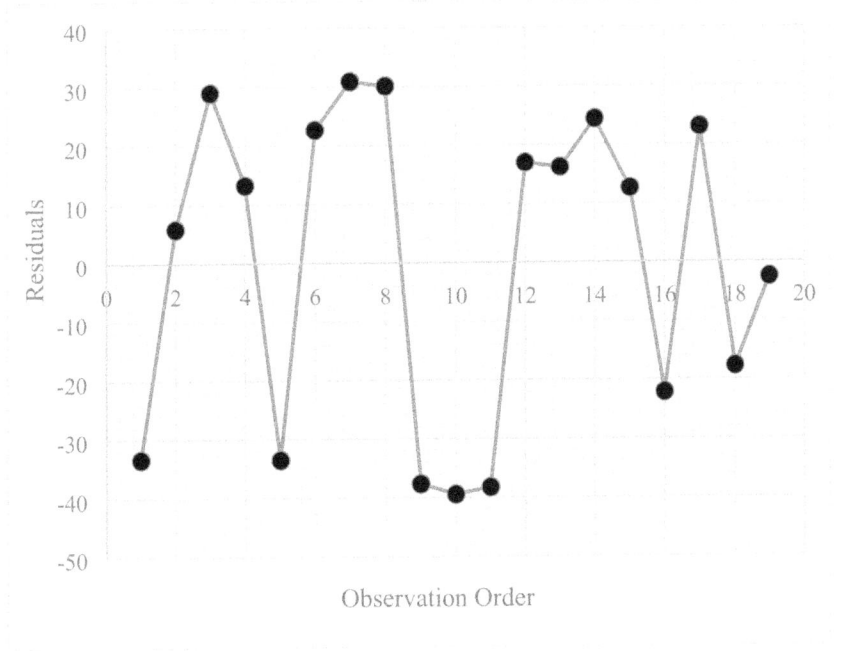

Fig. 2.6: Normal probability plot of residuals vs experimental order

(ii) Box & Whisker plot

Box & Whisker plots were developed by an American mathematician called John Wilder Tukey (Sande, 2001). Box & Whisker plot is used to demonstrate the shape of a distribution, its central value and its variability. The ends of a Box & Whisker plot, depict the lower and upper quartiles. The difference between lower and upper quartiles gives the interquartile range (IQR) and the median is shown as a vertical line inside the box. The vertical lines stretching from the box are called whiskers. Whiskers represent the variability outside the upper and lower quartiles. Outliers are plotted as individual points. In a perfectly normal distribution, the median is equal to mean and the mode and located in the middle of the interquartile range (Manikandan, 2011).

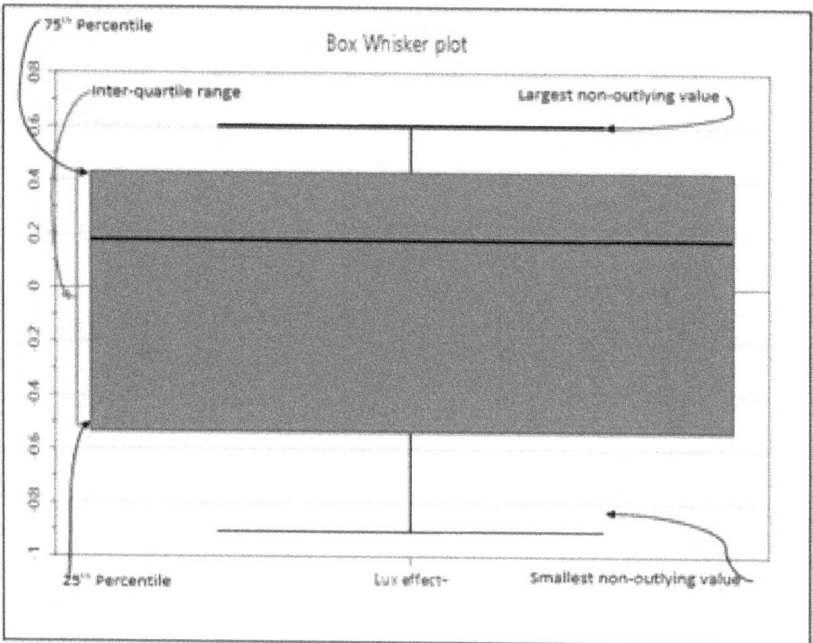

b (iii) Histogram

In general, there are different shapes that are possible in a histogram, i.e. normal distribution, skewed, double-peak (bimodal), plateau and edge peak. A random variable x follows a normal distribution if the probability;

$$f(x) = \frac{1}{\sigma\sqrt{2\pi}} e - \frac{1}{2}(\frac{x-\mu}{\sigma})^2$$

Where $-\infty < x < \infty$ (This is a bell-shaped curve)

Therefore, $x \sim N(\mu, \sigma)$, *meaning* x follows a normal distribution, with mean μ and standard deviation σ. The mean represents the center of the distribution and standard deviation represents the variation around the mean. In this example, the histogram of the data is;

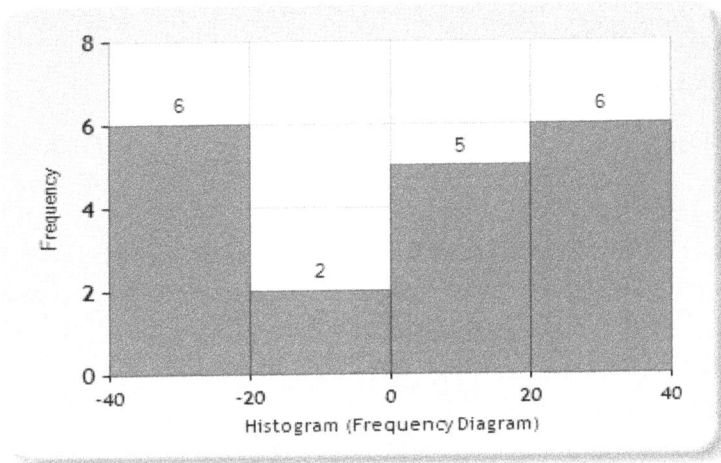

Fig. 2.8: Histogram plot of the response variable (laxative)

The observation in Fig. 2.7 (Box & Whisker plot) is confirmed by the histogram plot (Fig. 2.8). The histogram shows that the data is not normally distributed. A skew to the left is observed.

c) Theoretical Model

$$y = \beta_0 + \beta_1 x_1 + \beta_2 x_2 + \beta_3 x_3 + \beta_4 x_4 + \beta_{12} x_1 x_2$$
$$+ \beta_{13} x_1 x_3 + \beta_{14} x_1 x_4 + \beta_{23} x_2 x_3 + \beta_{24} x_2 x_4$$
$$+ \beta_{34} x_3 x_4 + \beta_{123} x_1 x_2 x_3 + \beta_{134} x_1 x_3 x_4$$
$$+ \beta_{1234} x_1 x_2 x_3 x_4 + \varepsilon$$

For this 2^4 full factorial experiment, we can fit a model containing a mean term, four (4) main effect terms, six (6) two-factor interactions, two (2) three-factor interactions and one (1) four-factor interactions. However, it is advisable to assume all three factor and higher interactions are not significant, based on sparsity-of-effects principle. Sparsity-of-effects principle is sometimes referred to as hierarchical ordering principle and states that it is likely that main and two-factor interactions are the most important in a factorial experiment (Wu, 2000). This means that higher order interactions such as three factor interactions are rare.

d) Fitting the Model to the data

The fit has a large R^2 (0.96) and adjusted R^2 (0.91). The ANOVA table shows that the regression model is significant ($P<0.05$). Lack-of-fit error is also significant ($P<0.05$). However, one main effect (ethyl alcohol) and two 2-factor interactions (SSA*SSH) and Chl*Eth are not statistically significant ($P>0.05$). The latter can be considered unnecessary terms in the model (Table 2.11).

Table 2.11: ANOVA table

	DF	Prob>F
Total	19	
Constant	1	0.000
Sodium sulphate anhydrate (SSA)		0.100

Sodium sulphate hydrate (SSH)		0.002
Chloroform (Chl)		0.002
Ethyl alcohol (Eth)		0.430
SSA*SSH		0.399
SSA*Chl		0.001
SSA*Eth		0.0035
SSH*Chl		0.0011
SSH*Eth		0.012
Chl*Eth		0.431
Total Corrected	18	
Regression	10	0.00
Residual	8	

The full regression model with both significant and non-significant factors is:

$$Laxative\ effect = 58.5 + 3.63 * SSA + 8.50 * SSH$$

$$+ 8.75 * Chl - 1.63 * Eth - 2.23 * SSA * SSH$$

$$- 10.13 * SSA * Chl - 8 * SSA * Eth$$

$$- 9.75 * SSH * Chl + 6.38 * SSH * Eth$$

$$- 1.63 * Chl * Eth$$

Based on ANOVA results, it is important to consider removing the non-significant variables with P-values greater than 0.05. However, it is equally important to be aware of the fact that removal of the non-significant main effects such as Sodium sulphate anhydrate (SSA) and sodium sulphate hydrate (SSH) has the effect of removing the significant interactions between these main effects in the model. For the purpose of this exercise, the non-significant main effects were not removed. However, non-significant interactions SSA*SSH and Chl*eth (P>0.05) were removed, following which the ANOVA was redone. . The removal of the non-significant factor interactions had the effect of lowering the R^2 and adjusted R^2 values as shown in Table 2.12:

Table 2.12: Reduced model after removal of non-significant two-factor interactions

Luxative Effect	Coefficient	P value

Constant	58.24	0.000
Sodium sulphate anhdyrate (SSA)	3.89	0.074
Sodium sulphate hydrate (SSH)	8.76	0.001
Chloroform (Chl)	9.01	0.001
Ethyl alcohol (Eth)	-1.36	0.490
SSA*Chl	-10.39	0.000
SSA*Eth	-8.26	0.002
SSH*Chl	-10.01	0.000
SSH*Eth	6.11	0.010
R^2	0.95	
R^2 adj	0.91	

e) **Testing the residual assumptions of the reduced model**
(i) *Residuals vs predicted response*

This is the most frequently plotted plot in residual analysis. This is a scatter plot of residuals on y-axis and fitted/predicted responses on the x-axis. The plot is used to test for non-linearity, unequal error variance and outliers. Points that are randomly scattered around the line imply that the model fits the data well, as shown in Fig. 2.9.

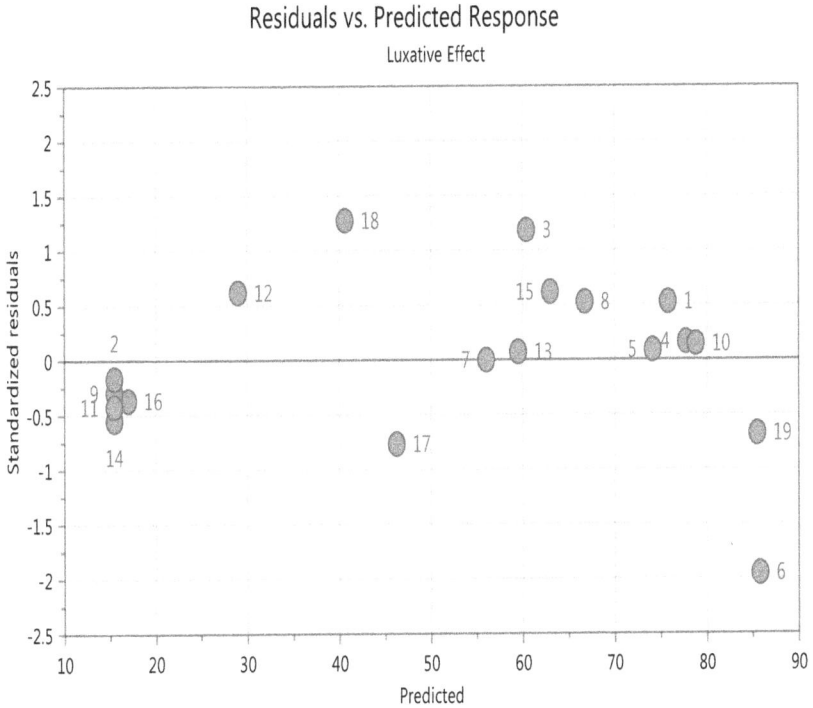

Fig. 2.9: Residuals vs predicted responses

ii)   *Normal probability plot of residuals*

Standardized residual are a measure of the strength of the difference between observed and expected values (Sharpe, 2015). In general, if the data is normally distributed, 95% of the data should lie within 2 standard deviations from the mean. Data that lies outside the 2 standard deviations is considered outlier. In this example, residuals are approximately normally distributed around the diagonal line, with no value greater than 2 standard deviations (Fig. 2.10).

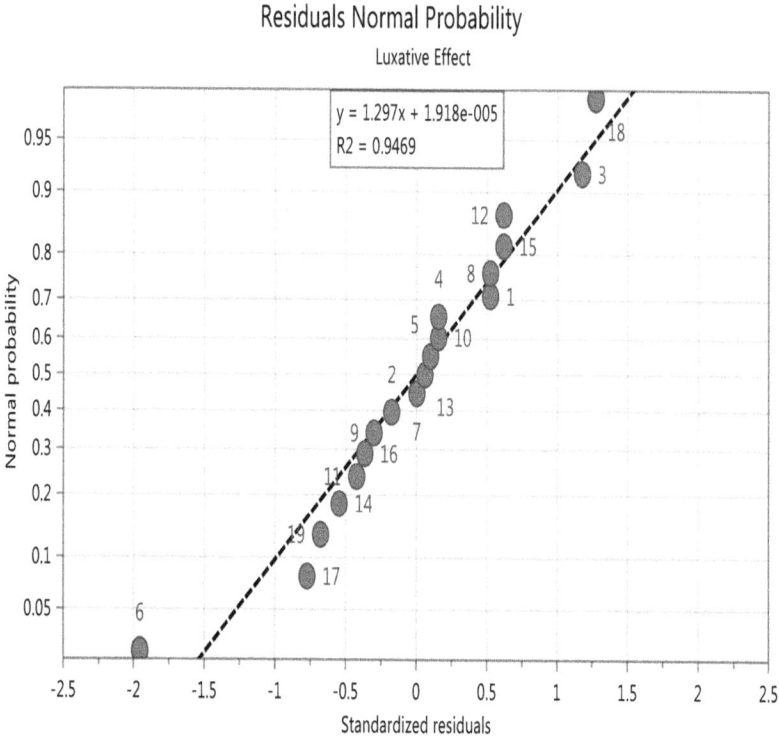

Fig. 2.10: Normal probability plot of residuals

iii) *Residuals vs experimental run order*

The plot shows the residuals plotted against order of experimental runs, in the two-level experimental design, in problem 3 above. Randomly distributed points indicate that the sequence of the experiments does not affect the residuals.

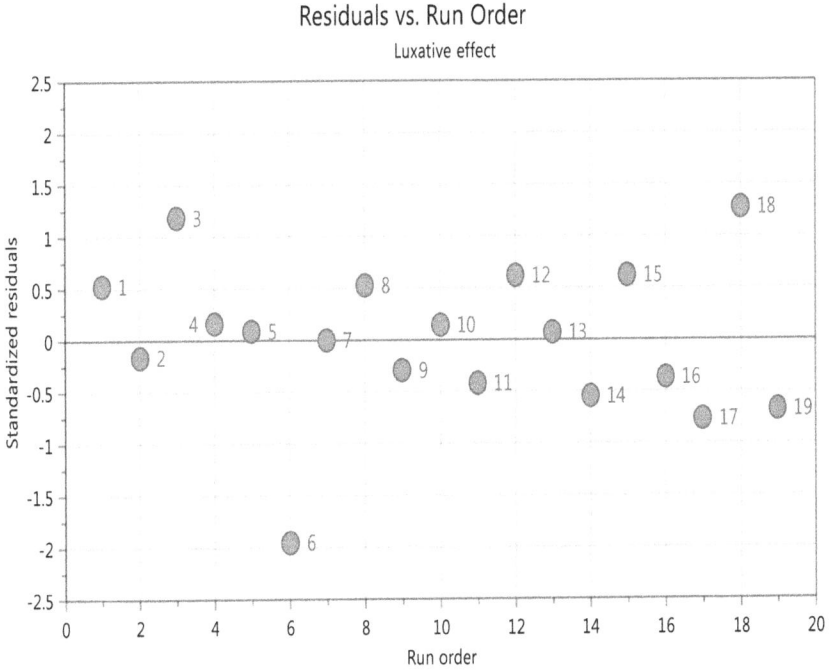

Fig. 2.11: Residuals vs run order

f) Main effects of the four laxative constituents

Main effect is the effect of a single independent variable on a dependent variable, ignoring all other independent variables.

Calculation of Main Effects based on Main Effects Plots:

Table 2.13: Main Effects of the four laxative constituents

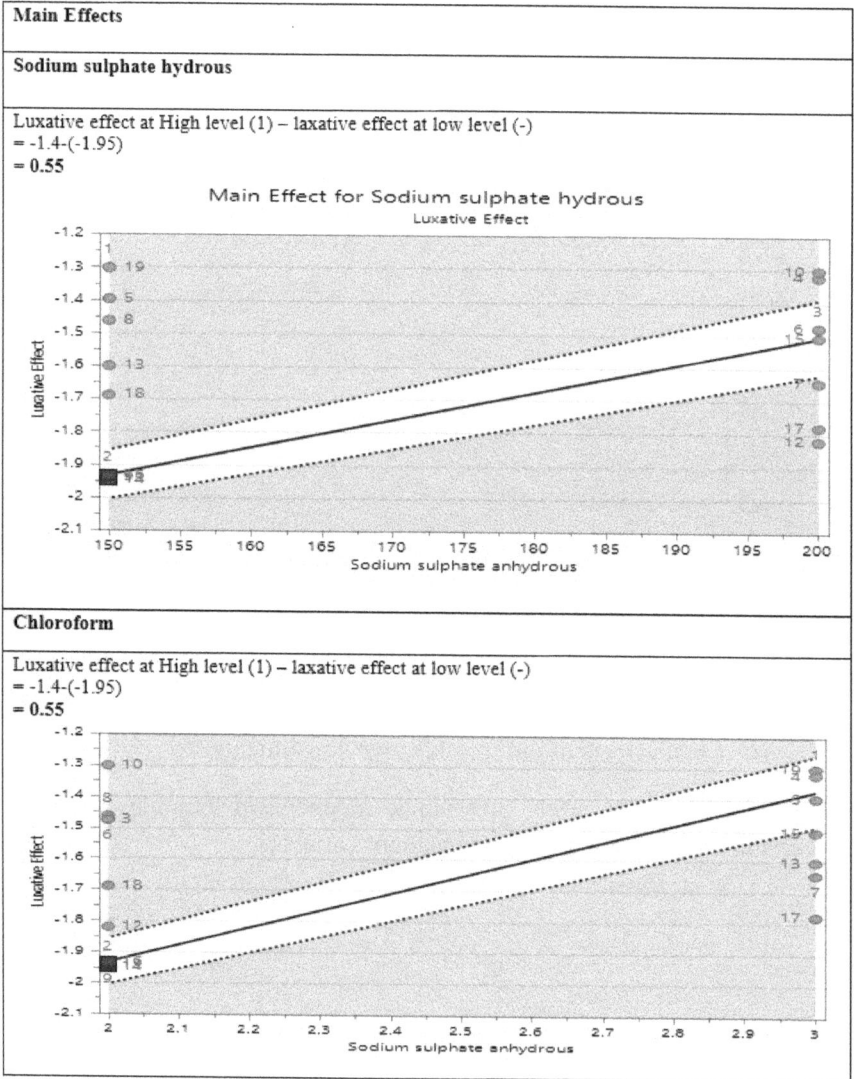

Main Effects
Sodium sulphate hydrous
Luxative effect at High level (1) – laxative effect at low level (-)   = -1.4-(-1.95)   = 0.55
Chloroform
Luxative effect at High level (1) – laxative effect at low level (-)   = -1.4-(-1.95)   = 0.55

Example 4: A car manufacturing company conducted a survey to determine changes in sales of their latest car based on the difficulty of maintenance (x_1) and cost of spare parts (x_2). The difficulty of maintenance and cost of spare parts where rated from low (-1) to high (+1) levels. The data is shown in Table 2.14, below.

Table 2.14: Calculation of main effects by HAND

X_1 (Difficulty of maintenance)	X_2 (Cost of spare parts)	Sales volume
-1	-1	80
+1	-1	60
-1	+1	45
+1	+1	30

Show the Calculation of Main Effects by Hand:

  a) Ease of maintenance (x_1),
  b) Cost of spare parts (x_2) and
  c) interaction effect (x_1x_2)

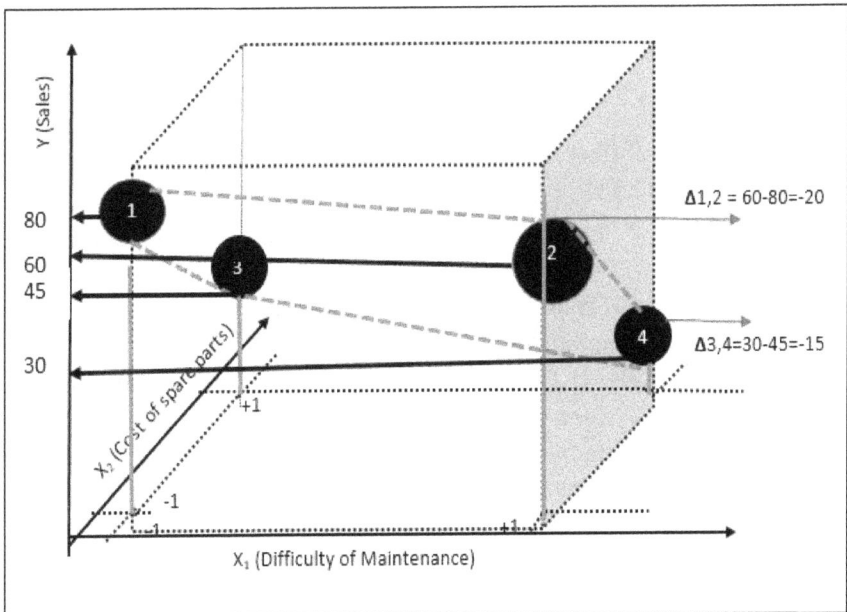

Fig. 2.11. Calculation of main effects of ease of maintenance and cost of spares by Hand

61

a)   Main effect of Difficulty of Maintenance

There are two estimates to calculate main effect of ease of maintenance from low (-1) to high (+1). The first estimate $\Delta1,2$, the change in Sales volumes at low cost of spare part is -20, shows that car sales decreased by 20%. The second estimate, $\Delta3,4$ indicates that the sales dropped by 15%. Therefore the main effect of ease of maintenance is the average of $\Delta1,2$ and $\Delta3,4$.

$$(\Delta1,2 \text{ and } \Delta3,4/2) = (-20 + -15)/2 = -17.5\%$$

The main effect of difficulty of maintenance is that as the difficulty of maintaining a car increased from low to high, a mean drop of 17.5% in car sales was observed.

b)   Main Effect of Cost of Spare parts

The main effect of cost of spares is the average of the estimates of $\Delta1,3$ and $\Delta2,4$

$$\Delta1,3 = 45 - 80 = -35 \text{ and } \Delta2,4 = 30 - 60 = -30$$

$$(\Delta1,3 \text{ and } \Delta2,4/)2 = (-35 + -30)/2 = -32.5\%$$

Therefore, the main effect of cost of spare parts on the sales volumes of the new car model is **-32.5%**

Example 5: The stability of a diarrhea mixture was evaluated based on the amount of a primary degradation product produced at the end of a 12 months period. Using a two-level full factorial design, 16 different formulations of the diarrhea mixture were developed. Each formulation was stored at the same temperature and humidity conditions. Four ingredients of the diarrhea mixture (kaolin, pectin, methyl hydroxybenzoate and propyl benzoate) were studied at two levels, low level (-1) and a high level (+1).

Table 2.15: Factor settings of the diarrhea mixture constituents

Name	Low Level		High Level	
Kaolin	1.500g	-1	2.000g	+1
Pectin	50mg	-1	60mg	+1
Methyl Hydroxybenzoate	0.2%	-1	0.4%	+1
Propyl Hydrobenzoate	0.02%	-1	0.04	+1

A two-level full factorial design with 4 factors (kaolin, pectin, methylhydroxybenzoate and propylbenzoate) and 1 response variable (amount of degradation product), gives a total number $2^4 = 16$ experimental runs.

Table 2.16: Two level full factorial design matrix showing 16 formulations and their corresponding amounts of degradation product

Experiment	Kaolin	Pectin	Methyl Hydroxybenzoate	Propyl Hydroxybenzoate	Response (Amount of Degradation product)
1	-1	-1	-1	-1	3.9
2	1	-1	-1	-1	4
3	-1	1	-1	-1	4.01
4	1	1	-1	-1	3.7
5	-1	-1	1	-1	3.62
6	1	-1	1	-1	3.35
7	-1	1	1	-1	4.28
8	1	1	1	-1	3.24
9	-1	-1	-1	1	3.2
10	1	-1	-1	1	3.62
11	-1	1	-1	1	3.5
12	1	1	-1	1	2.69
13	-1	-1	1	1	3.45
14	1	-1	1	1	3.06
15	-1	1	1	1	4.09

16	1	1	1	1	3.18

Question: Show using a two-level full factorial design

   a) Normal probability plot
   b) Box plot of the response
   c) Histogram plot of the response
   d) Create the theoretical model relating the ingredients to the response
   e) Fit the model to data
   f) Regression analysis using ANOVA
   g) Plot the main and interaction plots
   h) Response contour plot

Answers

   a) Normal probability plot

The points lie approximately on the diagonal line with small scatter. The model can be considered to fit the data well. This residuals are approximately normally distributed around the diagonal, with no value greater than 2 standard deviations. In other words, no evidence of an outlier.

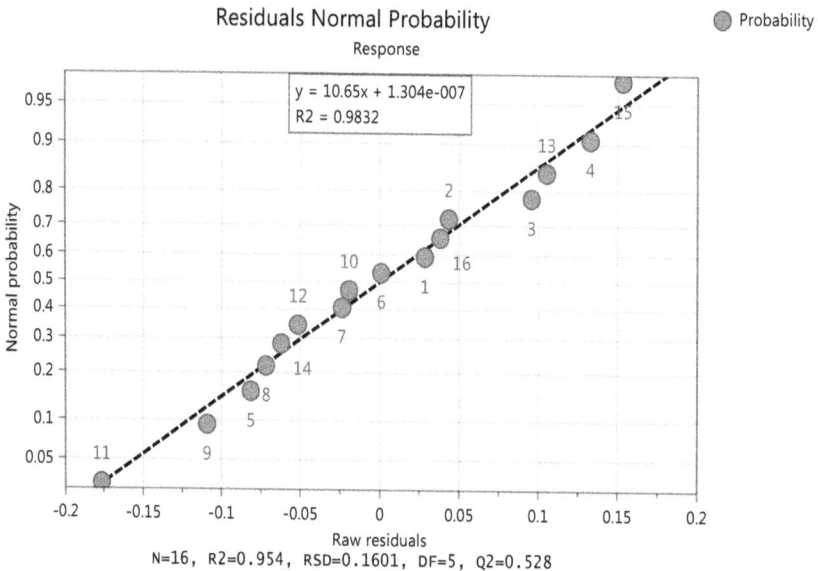

Residuals Normal Probability
Response

y = 10.65x + 1.304e-007
R2 = 0.9832

N=16, R2=0.954, RSD=0.1601, DF=5, Q2=0.528

Fig. 2.12: Normal probability plot of the residuals

b) Box & Whisker plot

The Box & whisker plot does not indicate any outliers in this response variable and the median is located in the middle of the 25th and 75th percentiles giving an idea of a normal distribution without a significant skew.

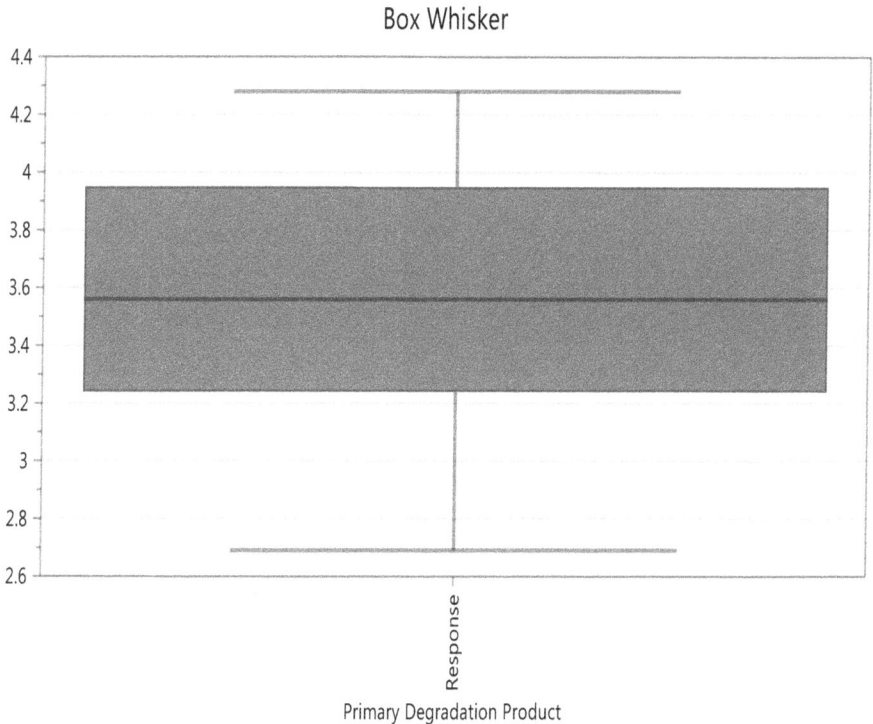

Box Whisker

Primary Degradation Product

Fig. 2.13: Box-Whisker plot

c) Histogram

A closer look at the histogram plot, suggested that a transformation could improve the fit of the data to our regression model, as shown in Fig. 2.14. In this example, the histogram Fig. 2.14 (i) has a slightly heavier tail to the right. Hence a logarithmic transformation can be used to correct this, as shown in Fig. 2.14 (ii). Log transformation is a widely used method for addressing skewed data and is one of the most popular transformations used in biomedical and psychosocial research (Changyong, 2014).

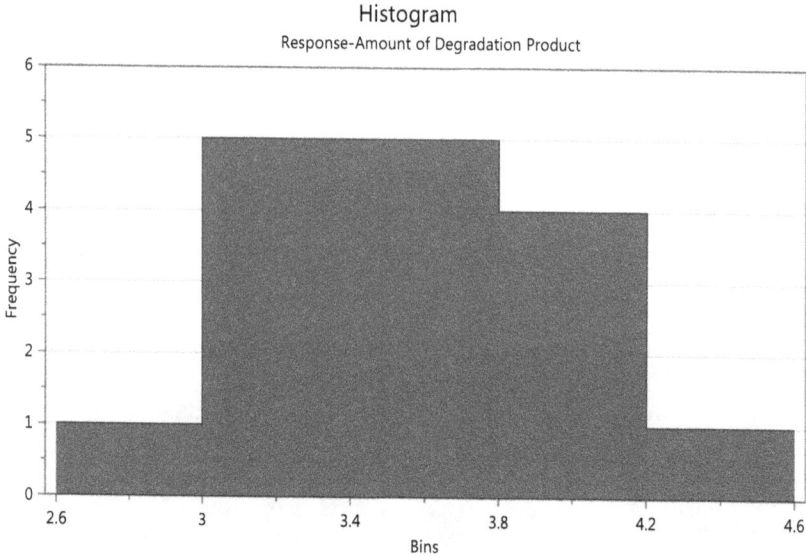

Histogram
Response-Amount of Degradation Product

(i)   Untransformed histogram

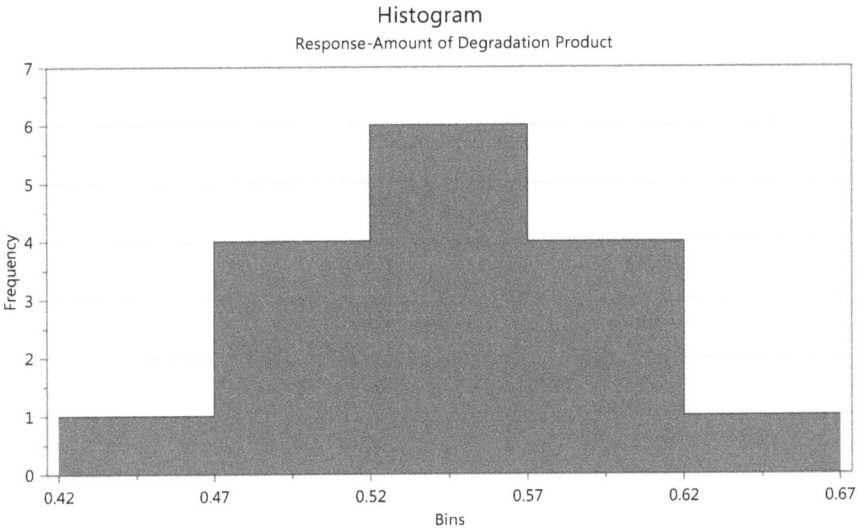

Histogram
Response-Amount of Degradation Product

(ii) Log transformed histogram

Fig. 2.14: Histogram plot of the response

In design of experiments, histograms provide clues about the need for transformation of responses. In this example, the histogram has a slightly heavier tail to the right. A logarithmic transformation was used to correct this. Log transformation reduces variability of data, especially in data with outliers.

d) Create the theoretical model

For this four factor, 2 level full factorial experiment, we can fit a model containing a mean term, four main effect terms, six two-factor interactions, two three-factor interactions and one four-factor interactions. All interactions of three or more factors were considered not to be statistically significant, based on sparsity-of-effect principle.

$$y = \beta_0 + \beta_1 x_1 + \beta_2 x_2 + \beta_3 x_3 + \ldots + \varepsilon$$

e) Fitting a full factorial design model to the DoE data, gave the following summary statistics;

- R-Square = 0.93, R-Square adjusted = 0.78, Number of Observations = 16
- Confidence level = 95%

Table 2.17: Coefficients for the model

Response (Amount of Degradation Product)	Coefficient	Prob.>F
Constant	0.548	0.000
Kaolin	-0.025	0.009
Pectin	0.002	0.719
Methyl Hydroxybenzoate	-0.002	0.707
Propyl Hydrobenzoate	-0.026	0.008
Kao*Pec	-0.022	0.014
Kao*Met	-0.014	0.064
Kao*Pro	-0.003	0.673
Pec*Met	0.016	0.044
Pec*Pro	-0.002	0.727
Met*Pro	0.015	0.058

The fit has a large R^2 and adjusted R^2 of 0.93 and 0.78 respectively. However, there is a large number of unnecessary terms with high number of P>0.05 values (P-values). Therefore it is important to consider removing the non-significant variables with P-values greater than 0.05. Two main effects (Pectin and methyl hydroxybenzoate) are not significant. However, interactions with other main factors are significant e.g. kaolin & pectin ($P<0.05$) and pectin & propyl hydroxybenzoate ($P<0.005$). Therefore, it is not wise to remove the non-significant main factors to avoid changing the hierarchy of the model.

f) Regression analysis using Analysis of Variance (ANOVA)

Using ANOVA, it is possible to assess the significance of the regression model and lack-of-fit of the model. The model is significant (P<0.005). In this example, no experimental runs were repeated, hence the comparison between model error and replicate error, referred to as lack of fit did not yield any results.

Table 2.18: Regression analysis using an ANOVA

	DF	Sum of Squares (SS)	Mean Square	F-ratio	P-value
Total	16	4.84552	0.30285		
Constant	1	4.80221	4.80221		
Total Corrected	15	0.04330	0.00289		
Regression	10	0.04034	0.00403	6.81128	0.024
Residual	8	0.00296	0.00059		
Lack-of-fit	5			-	
Pure Error	3				

g)   Plot the main and interaction plots

    i) Main Effects

69

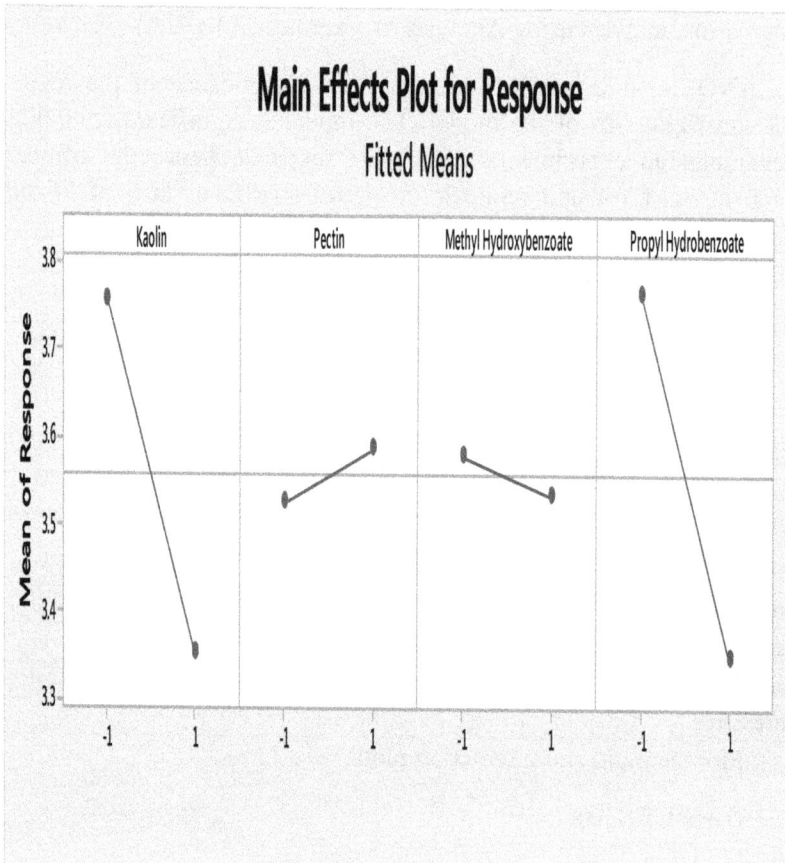

Fig. 2.15: Main effects of the laxative ingredients

The main effect of a factor is the change in the response variable (amount of degradation product) as a result of changing any one factor from its low (-1) to (+1) level, while keeping the other factors at their average values. The main effect of kaolin is about 37.5mg-33.5mg = 4mg. There is no change in the amount of the degradation product as pectin and methyl hydroxybenzoate are varied from their low (-1) to high (+1) levels.

The main effect of propyl hydroxybenzoate is 37.5mg-33.3mg = 4.2mg. It is however, helpful to accompany main effect analysis with an ANOVA analysis in order to evaluate the statistical significance of the changes in the main factors.

In addition, if the interactions are significant, you may not interpret main effects without paying attention to the interaction effects.

In this example, kaolin and propyl hydroxybenzoate are associated with the greatest amounts of degradation products at the end of the 12 months period. However, the two-way ANOVA analysis indicates that these main effects are statistically significant. It can be concluded that as the levels of kaolin and propyl hydroxybezoate change from low (-1) to high (+1), there was a significant drop in the amount of the degradation product.

ii) Interaction plots

Interaction refers to the situation where the effect of one factor depends on the level of another factor. Parallel lines in an interaction plot indicate that there is no interaction between factors. The greater the difference in slope between the lines, the higher the degree of interaction.

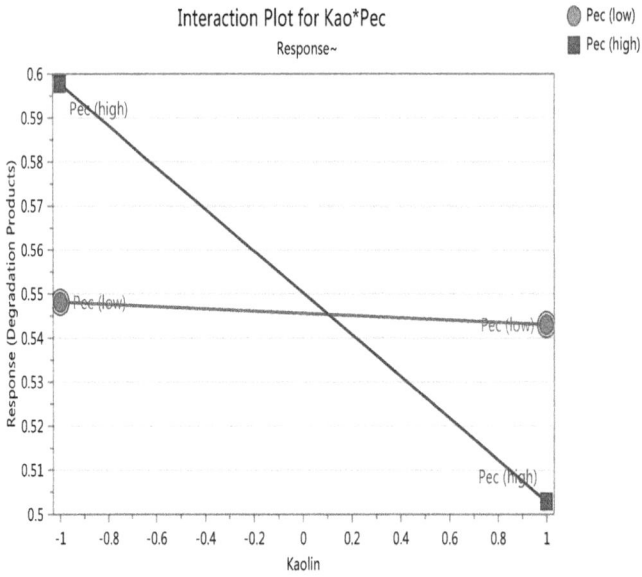

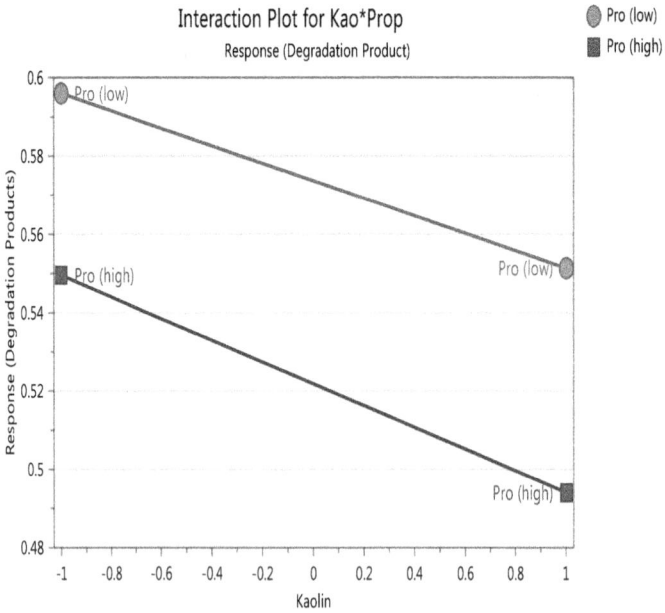

Fig. 2.16: Interaction effects of laxative ingredients

Increasing the amount of kaolin from low (-1) to high (+1) and keeping pectin at low (-1) level results in a steep decrease in the amount of degradation product at the end of a 12 months period. However, an increase in kaolin from low (-1) to high (+1) at high levels (+1) of pectin results in a slight change in the amount of degradation product. This indicates that there is a strong interaction between pectin and kaolin, which means that the effect one factor has on the response is dependent on the level set for the factors. However, there is almost no interaction between kaolin and propyl hydroxybenzoate.

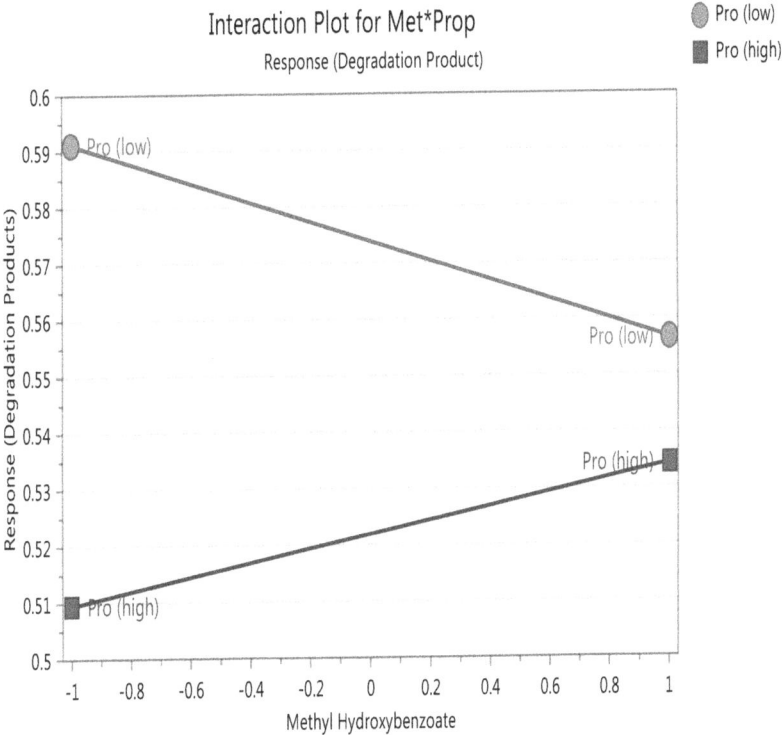

Fig. 2.17: Interaction effects of laxative ingredients

The graph shows a mild interaction between methyl hydroxybenzoate and propyl hydroxyl benzoate. However, it is important to note that interaction plot does not provide information on whether the interaction is statistically significant or not. A summary of all interactions is shown in Fig. 2.18.

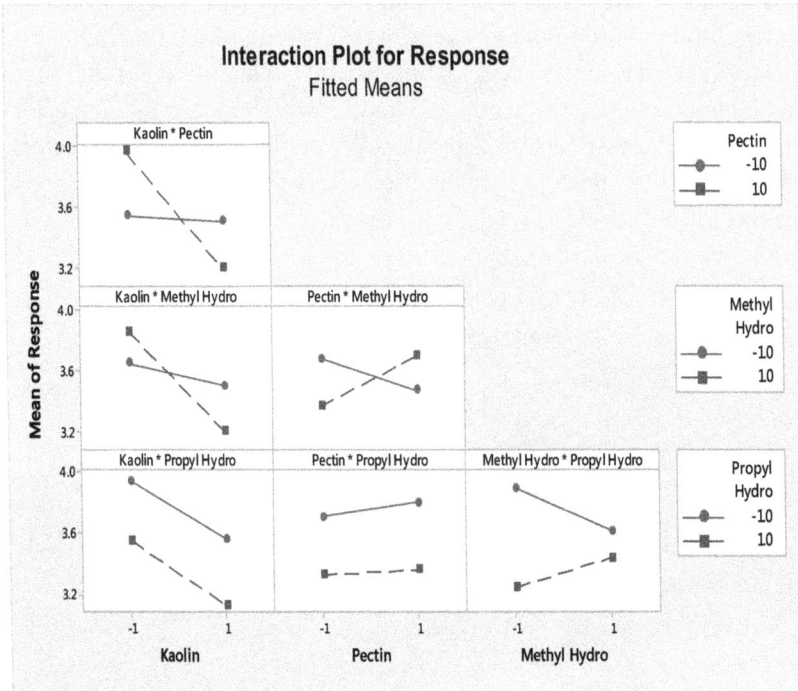

Fig. 2.18: All two-factor interactions

h) Response Contour Plot

A contour plot is a graphical representation of the relationship between three numerical variables in two dimensions. Two variables are x and y and the third is the response variable. Therefore, a contour plot enables the exploration of the potential relationship between two independent variables and one dependent variable. DOE contour plots assist in the determination of settings that will either minimize or maximize the response variable. In addition, DoE contour plots assist in determining the experimental settings that result in obtaining a desired value of the response variable. In this example, a decrease in kaolin towards the low level (-1) and an increase in pectin from low (-1) to high (+1) results in an increase in the amount of degradation product at the end of 12 months period.

74

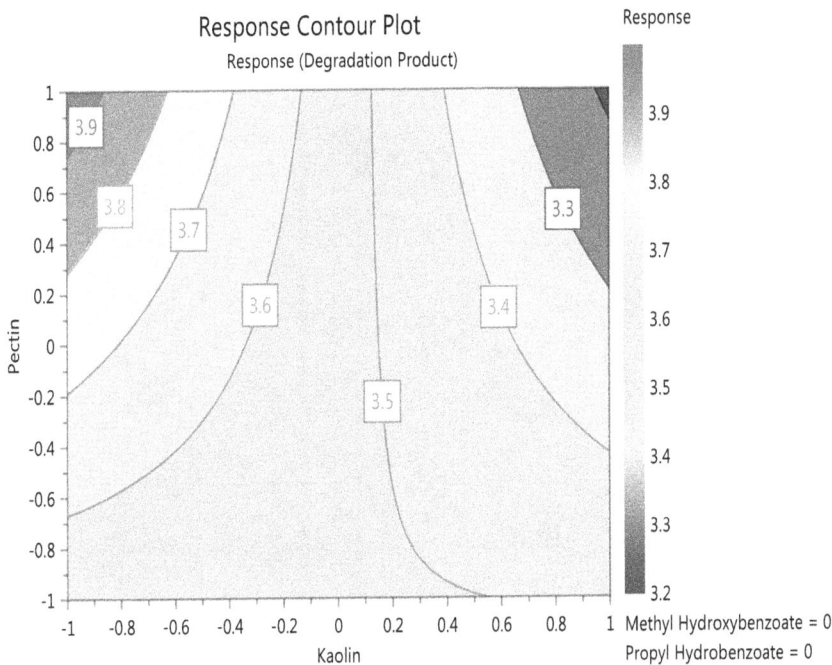

Fig. 2.19: Response contour plot

g) Pareto Plot

The purpose of the Pareto chart is to distinguish between a few important and many unimportant factors. Pareto charts are based on a Pareto principle that 20% of the work can achieve 80% of the benefits of doing the entire job. In Fig. 2.21, the horizontal line bars of the Pareto chart show the amount of degradation product. Each horizontal line represents the contribution to the total by each factor. The bars are arranged in a ranked order, with respect to their contribution to the degradation product. Therefore, Fig. 2.21 shows that main effects propyl hydroxybenzoate and kaolin contribute to the 80% of degradation.

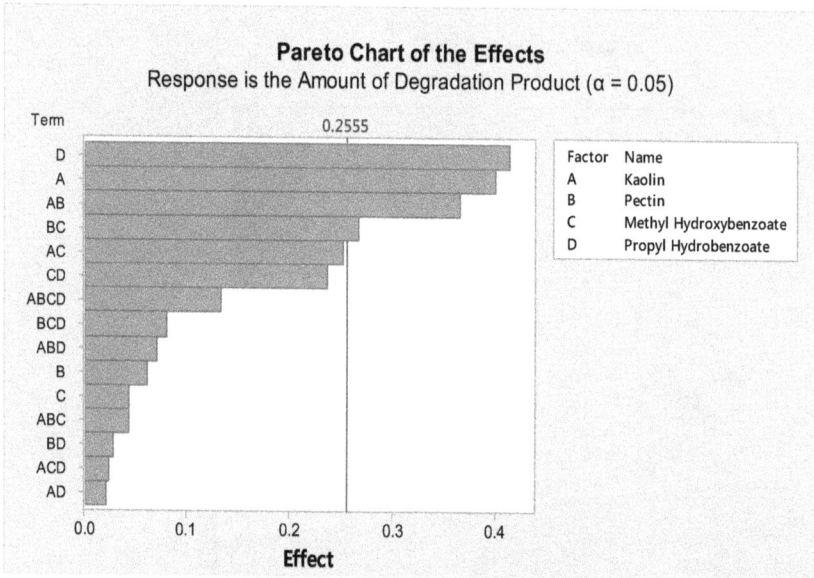

Fig. 2.20: Pareto Chart

References

1. Peng, C.Y.J., Lee, K.L., & Ingersoll, G.M. (2002). An introduction to logistic regression analysis and reporting. The journal of educational research.

2. Sibanda, Wilbert, Philip Pretorius & Anne Grobler. Response surface modeling and optimization to elucidate the differential effects of demographic characteristics on risk of HIV infection in South Africa. Advances in social network analysis and mining (ASONAM), 2012 IEEE/ACM, 2012.

3. Games, P.A., Winkler, H.B., & Probert, D.A. (1972) Robust tests for homogeneity of variance. Educational and Psychological Measurement, 32 (4), 887-909.

4. Carroll, R.J., & Schneider, H. (1985). A note on Levene's tests for equality of variances. Statistics and Probability Letters, 3(4), 191-194.

5. Altman, D.G. (1991). Randomization. BMJ: British Medical Journal, 302 (6791), 1481.

6. Sande, G. (2001). John Wilder Tukey. Physics Today, 54 (7), 80-81.

7. Manikandan, S. (2011). Measures of central tendency: Median and mode. Journal of Pharmacology and Pharmacotherapeutics, 2(3), 214.
8. Wu, C.F. Jeff; Hamada, Michael (2000). Experiments: Planning, analysis and parameter design optimization. New York: Wiley. P.112. ISBN 0-471-25511-4.
9. Sharpe, D. (2015). Your Chi-square test is statistically significant. Now what? Practical assessment, Research & Evaluation, 20.
10. Changyong, F.E.N.G., Hongyue, W.A.N.G., Naiji, L.U., (2014). Log-transformation and its implications for data analysis. Shanghai archives of psychiatry, 26(2), 105.
11. Bullock, D., & Urbanik, T. (1999). Traffic signal systems addressing diverse technologies and complex user needs. A3-A8: committee on traffic signal systems, 12.

Inspirational Inventions

Garrett Morgan

Garrett Morgan an African American inventor was born in Kentucky on March 4, 1877. Garrett Morgan had only an elementary school education and started his career as a sewing machine mechanic. After witnessing a collision between an automobile and a horse-drawn carriage, Morgan invented a traffic signal that was inexpensive. In 1923, Garrett Morgan patented the Morgan Traffic Signal, which he later sold to General Electric (Bullock, 1999). He also patented several other inventions such as improved sewing machine, a hair-straightening product, and a respiratory device that would later provide the blueprint for WWI gas masks.

Chapter 3: Analysis of DoE data

This chapter introduces the basic principles of the analysis of DoE data. This chapter will also describe the stages involved in DoE data analysis such as how to evaluate raw data, development of regression models and their interpretation. The end of this chapter, will be dedicated to the use of a regression model for prediction.

The analysis of data using DoE can be broadly grouped into three fundamental classes namely *raw data analysis*, *regression model analysis* and *how to use regression models for prediction*.

Raw Data Analysis

Data analysis is the process of providing order, structure and meaning to experimental data. Data is analyzed using descriptive and inferential statistics. Therefore data analysis seeks to transform raw data into useful information. Some of the methods used to analyze raw data include histograms, box & whisker plots and response vs scatter plots. Raw data analysis also aims to examine outliers, typological mistakes and some problems relating to raw data.

Demonstration of Raw Data Analysis

In a **Case-Study 1**, using a Plackett-Burman design to develop and optimize a novel cross-linked calcium-aluminum-alginate-pectinate gelisphere system for in-vitro site-specific release of Mentha piperita, an essential oil for the treatment of irritable bowel syndrome, 17 formulations were developed and four responses recorded. Two of the four responses studied were unhydrated matrix resilience and hydrated matrix resilience (Sibanda, 2004).

To illustrate the analysis of raw data for a DoE study, only two responses from the Plackett-Burman design will be studied, namely unhydrated matrix resilience and hydrated matrix resilience at pH 3.

Table 3.1: The Plackett-Burman Design matrix

Oilisphere formulation	Unhydrated matrix resilience	Hydrated matrix resilience (pH 3)
F1	8.48	3.98
F2	3.08	4.27
F3	7.97	23.93
F4	0	0
F5	102.01	4.47
F6	71.02	4.23
F7	0	0
F8	47.64	23.23
F9	0	0
F10	53.72	34.64
F11	53.72	34.64
F12	0	0
F13	94.95	67.76
F14	19.65	4.77

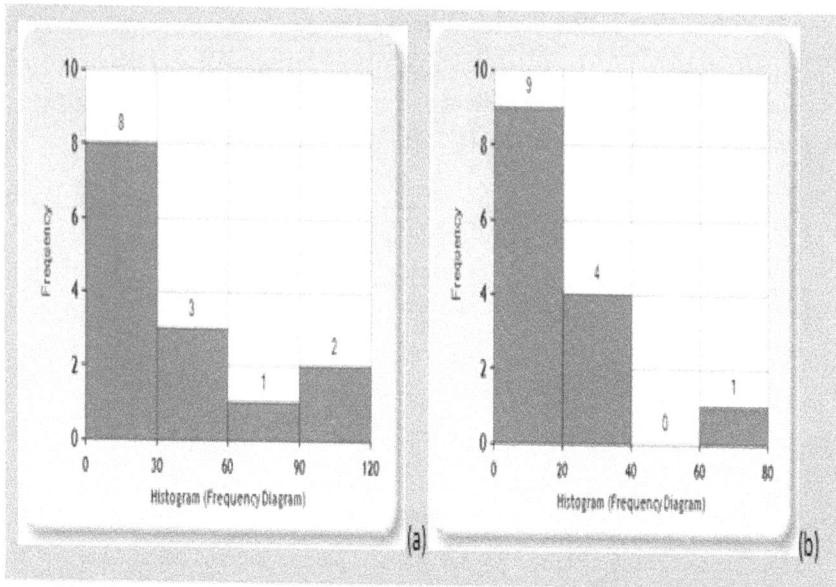

Fig. 3.1: Histograms of (a) Unhydrated Matrix Resilience (b) Hydrated Matrix Resilience

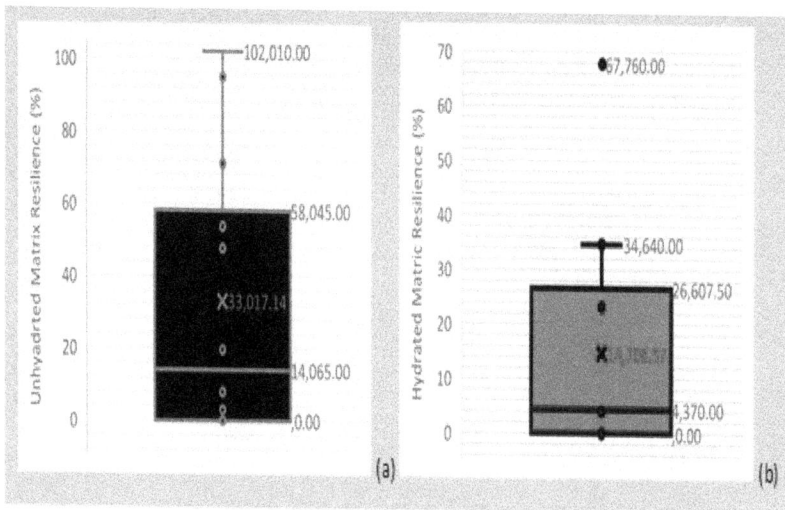

Fig. 3.2: Box-Whisker plots of a) Unhydrated Matrix Resilience and b) Hydrated Matrix Resilience

Unhydrated Matrix Resilience

Unhydrated Matrix Resilience

Hydrated Matrix Resilience

(a)

(b)

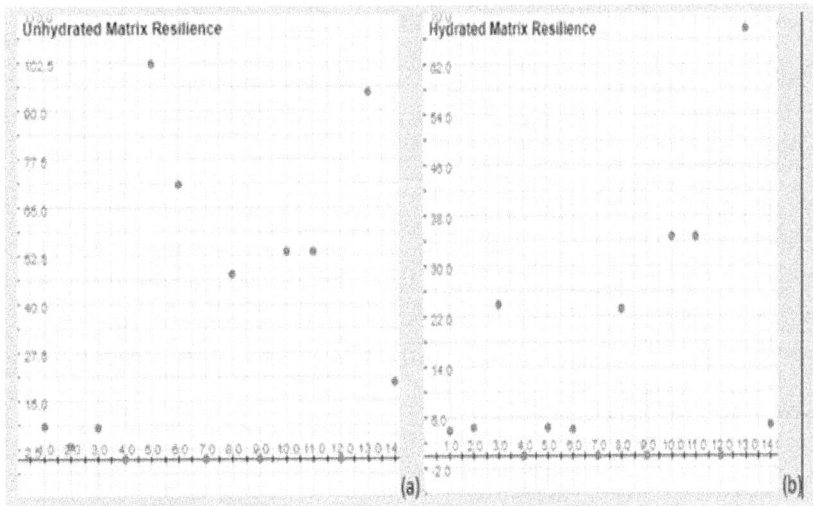

Fig.3.3: Response vs experiment run scatter plots a) Unhydrated Matrix Resilience and b) Hydrated Matrix Resilience

3.1. Residual Analysis

Histograms (Fig. 3.1) provide visual interpretation of numerical data and the range of values are called classes or bins. The frequency of data in each class is shown by the use of bars and the higher the bars, the greater, the frequency of data values in that bin.

Box and Whisker plots (Fig. 3.2) assist in determining the presence of outliers in the data. Interquartile range (IQR) is the difference between the first (Q_1) and third (Q_3) quartiles. IQR is a measure of spread and variability in a dataset. An outlier is an observation that falls more than $1.5 * IQR$ above the third quartile or below the first quartile. Therefore, a suspected low outlier is any value $< Q_1 - (1.5 * IQR)$ and a high outlier, any value $> Q_3 + (1.5 * IQR)$. Box and whisker plots also indicate if the data is normally distributed. A skew to the right is observed for both unhydrated and hydrated matrix resilience responses.

A scatter plot (Fig. 3.3) shows the relationship or association between two quantitative variables. Therefore, a scatter plot provides information on whether the independent and dependent variables are related.

3.2. Regression model analysis and evaluation of regression model fit

A regression analysis provides information on the relationship between response and predictor variables. Furthermore, regression analysis can be used for the purposes of model prediction and forecasting.

A regression model with a good fit has the advantage of generating predicted response values that are very close to the observed values. In other words, the error of prediction is drastically reduced. Model validation is therefore paramount to the development of a regression model with good fit. Model validation tools include numerical and graphical residual analysis techniques.

Numerical model-fit measures

(i)   R²

R^2 is one of the most important numerical measures of Goodness-of-fit. R^2 is a measure of how close to the data the fitted regression line is. It is an indication of the variation in the dataset that is accounted for by the regression model. R^2 values range from 0 to 1, where on depicts a model that is unable to account for any variability in the response data around the mean. On the other hand, 1 indicates that the model is capable of explaining most if not all of the variability around the mean. The formula for R-squared is:

$$R^2 = 1 - \frac{SS_{Regression}}{SS_{Total}} = 1 - \frac{Sum\ Squared\ Regression\ Error}{Sum\ Squares\ Total\ Error} \quad (i)$$

$$SS_{Total} = \sum (y_i - \bar{y})^2 \quad (ii)$$

where SS_{Total} *is Sum of Squared Total Error,* y_i *is each data point,* $\bar{y}$ *is mean value*

$$SS_{Regression} = \sum (y_i - y_{Regression})^2 \quad (iii)$$

where $SS_{Regression}$ *is Sum of Squared Regression Error,* y_i *is each data point,* $y_{Regression}$ *is Regression value*

Ideally, the regression error should be zero. Therefore, R^2 is the percent of variance explained. This means that R^2 is the fraction by which the variance of the errors is less than the variance of the dependent variable.

Weaknesses of the R² measure

R^2 measure is unable to show if the coefficient estimates and predictions are biased. Therefore in order to determine coefficient and predictions bias, the experimenter needs to conduct a residual analysis study. A high R^2 measure does not necessarily demonstrate a good model. It is possible to have a high R^2 value for a bad model and a low R^2 for a good model,

(ii)   Q²

Q^2 is a measure of regression model prediction goodness-of-fit and is determined as follows;

$$Q^2 = 1 - \frac{\sum_{i}^{n}(Y_i - \hat{Y}_{i,i})^2}{\sum_{i}^{n}(Y_i - \bar{Y})^2} = \frac{Predicted\ sum\ of\ squares\ (PRESS)}{Total\ sum\ of\ squares} \qquad (ii)$$

PRESS is the prediction sum of squares and represents the sum of squares of the prediction error. It is used for the assessment of the predictive capability of the regression model. In general smaller PRESS values are more desirable and indicate a model with a good predictive ability. PRESS is also used to compute R^2 measures of models. Generally Q^2 values range between $-\infty$ and 1 and $Q^2 > 0.5$ means the model is good.

(iii)   Reproducibility

In general, reproducibility is defined as the ease with which an experiment or study can be replicated by an experimenter or any other person working independently. It is a measure of the closeness of the results from two or more experiments carried out by the same experimenter or different experimenters.

$$Reproducibility = \frac{Mean\ Square_{pure\ error}}{Mean\ Square_{total\ corrected}} \qquad (iii)$$

As a general guide, reproducibility values below 0.5 indicate increased pure error and this could imply a poor control of the experimental procedure.

3.3. Graphical Residual Analysis

$$\hat{\varepsilon}_i = y_i + \hat{y}_i, \quad i = 1,2,...,n \quad (iv)$$

Where y_i is observed y

$\hat{y}_i$ =fitted y

If the model contains an intercept, the sum of residuals and their mean is zero.

$$\sum_{i=1}^{n} \hat{\varepsilon}_i = 0, and\ \bar{\hat{\varepsilon}} = 0$$

Therefore, residuals represent the difference between observed and predicted response values and are used to demonstrate variation that is not explained by the fitted regression model. In general, residual plots are generally used to determine the quality and integrity of the regression model, through the evaluation of the underlying statistical analysis assumptions about residuals such as constancy of variance of errors over time, predictions and independent variables, independence of errors and normality of error distribution. Therefore if the assumptions are met, the residuals are expected to vary randomly around zero, with the same spread throughout the plot.

3.3.1. Evaluating Error Variance

Variance of the errors σ^2 indicates how much observations deviate from the fitted curve. This means that if σ^2 is small, the parameters, $\beta_0, \beta_1, ..., \beta_k$ will be reliably estimated and predictions $\hat{y}$ will also be reliable.

Frequently, the value of σ^2 is unknown, can be estimated from the sample as:

$$\hat{\sigma}^2 = S^2 = \frac{SSE}{n - number\ of\ parameters\ in\ model} = \frac{SSE}{n - (k+1)} = \frac{\sum_i (y_i - \hat{y})^2}{n - (k+1)}$$

Where $\hat{y}_i = b_0 + b_1 x_{1i} + b_2 x_{2i} + ... + b_k x_{ki}$

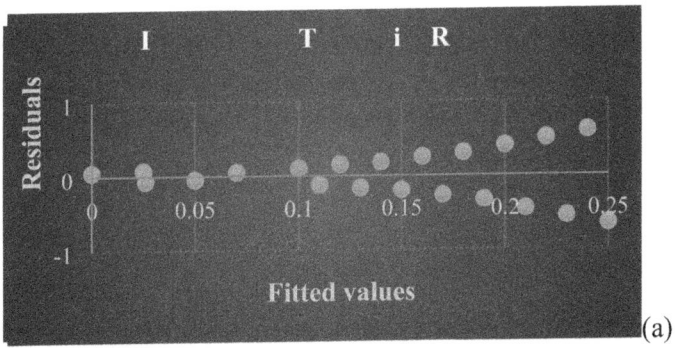

(a)

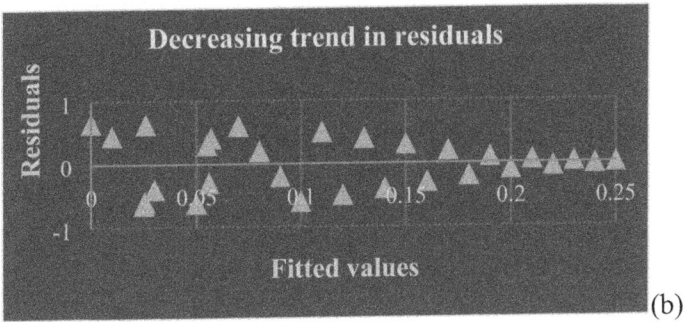

Decreasing trend in residuals

(b)

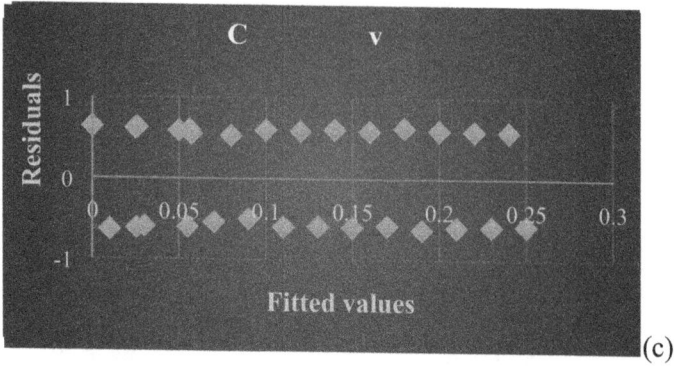

(c)

Fig 3.4. Error variance

In plots (*a*) and (*b*), the residuals are observed to increase and decrease with predictor variables respectively. It can be concluded from plots (*a*) and (*b*) that the assumption of constant variance is flouted, indicating that the regression model is not good.

3.3.2. *Independence of residuals over time*

In general, checking of independence assumptions focusses on error terms not outcome variables. Given an equation, $y = \alpha + \beta x + \varepsilon$, it means that y is the outcome variable and ε represents error terms.

Independence of residuals assumes that error terms are independent. The residual ε is represented by the difference between observed y_i and fitted $\hat{y}_i$ values. To check the independence of residuals, the residuals are plotted against time variables such as order of observation.

A pattern that is not random indicates lack of independence.

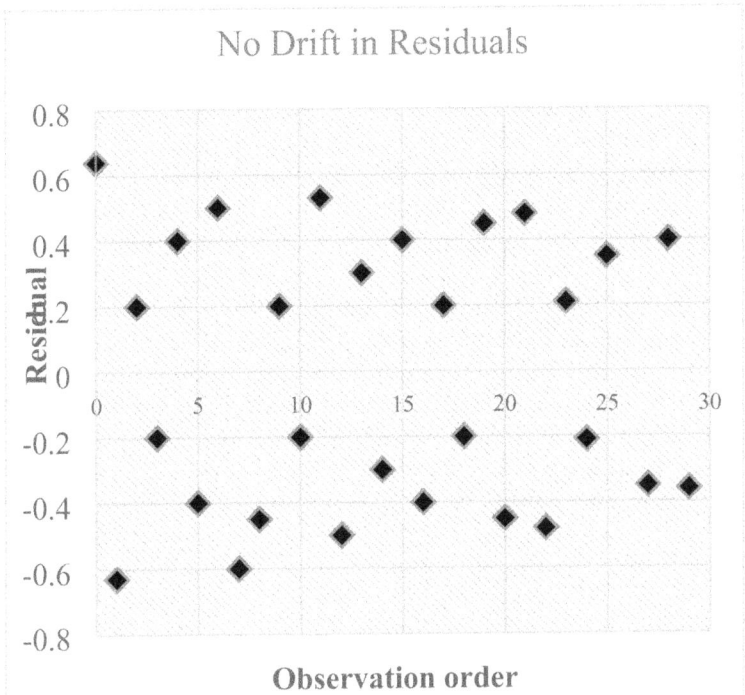

(a)

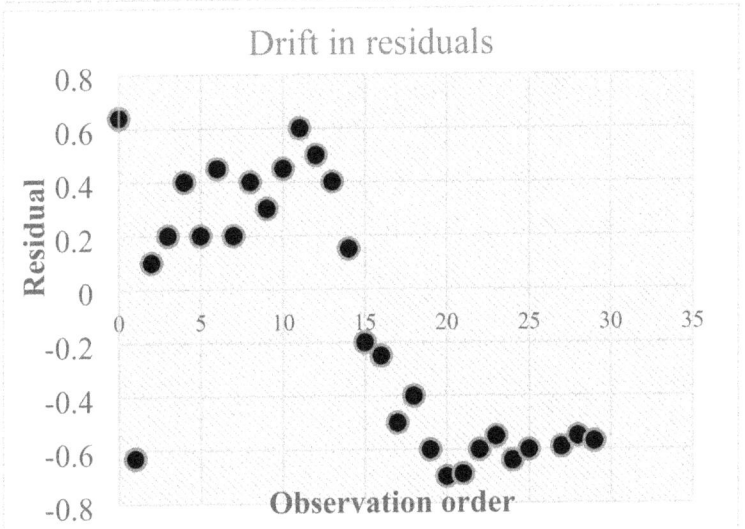

(b)

Fig. 3.5. Independence of residuals over time

In plot (a), residuals are observed to be randomly distributed around zero, indicating that there is no **drift in the residuals.**

3.4. Investigation of normality of error distribution

3.4.1. Histogram of residuals

Frequently, histogram plots are used to provide information on whether residual variance is normally distributed or not.

However, a histogram is not the best method for checking normality, because histograms of the same data with different bin size may look different. Therefore, it is more advisable to use a probability plot (also called a quantile plot or Q-Q plot). However, probability plots can be misleading for small data.

Example: A study was conducted to determine the ages of participants in a

Subject	Age (years)
1	36
2	25
3	38
4	46
5	55
6	68
7	72
8	55
9	36
10	38
11	67
12	45
13	22
14	48
15	91
16	46
17	52
18	61
19	58
20	55

(a)

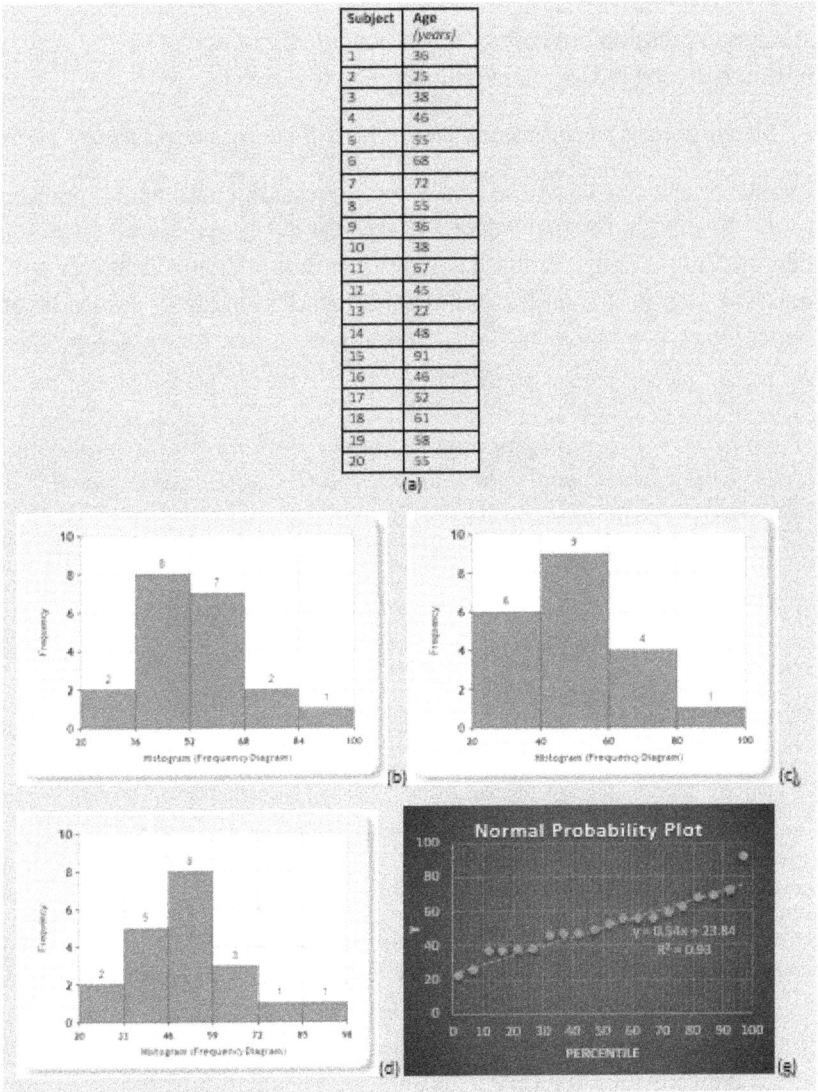

class. The results are shown below.

Fig. 3.6: a) Research data, Histogram plots with b) 4, c) 5 and d) 6 bins respective, (e) Normal probability plot

3.5. Model Interpretation

A regression expression describes the mathematical and statistical relationship between independent and dependent variables.

3.5.1. Interpretation of regression coefficients for linear regression

As explained earlier in this hand-book, the regression coefficients represent the average change in the dependent variable for every single unit change in the independent variable while keeping the other independent variables constant. The regression coefficient can be regarded as the measure of the importance of model terms and thus can be used to fine-tune the regression model.

In a study, using a two-level fractional factorial design to investigate the relationship between demographic characteristics and the risk of HIV infection amongst antenatal clinic attendees in South Africa.

$$Risk\ of\ HIV\ infectio = \quad 0.31 + 0.28 * woman's\ age$$

$$+ 0.08 * male\ partner's\ age$$

$$- 0.05 * Gravidity$$

$$- 0.11 * Parity$$

$$- 0.14 * \text{woman's educational level (vi)}$$

The equation (Eqn. vi) shows that the coefficient for the mother's age is 0.28. The coefficient indicates that for every additional unit change in mother's age, an increase of 0.28 in risk of HIV infection is expected. In addition, the equation shows that the coefficient for parity is -0.11. The coefficient indicates that for very additional unit change in parity, a decrease of 0.11 in risk of HIV infection is expected.

The constant by itself is meaningless as the demographic characteristics cannot be zero. However, the constant has to be included in the regression model for the following reasons;

i.   The regression model constant ensures that the residuals have a mean of zero,

ii. Exclusion of the regression model constant forces the regression model to pass through zero and thus presupposing that all the independent and dependent variables at that point are equal to zero.

3.6. How to use regression models for prediction

In situations where the regression model fits the data well, the regression model can be used to predict the value of one variable given the value of another variable.

Using *Equation (v)* above, the risks of HIV infection for women of different age-groups are shown below. (*To illustrate the effects of women and her male sexual partner's ages on the risk of HIV infection, gravidity, educational level and parity are kept constant*)

i.   *Young women sexually involved with older males*

Table 3.2: Risk of HIV infection amongst younger women

Mother's age (years)	Father's age (years)	Gravidity	Parity	Education (Grades)
15	65	9	8	10

$$Risk\ of\ HIV\ infection = 0.31 + 0.28(15)$$

$$+\ 0.08(65) - 0.05(9)$$

$$-\ 0.11(8) - 0.14(10)$$

$$=\ 6.98 \quad \text{(i)}$$

ii. *Older women sexually involved with same age or older males*

Table 3.3: Risk of HIV infection amongst older women

Mother's age (years)	Father's age (years)	Gravidity	Parity	Education (Grades)
45	65	9	8	10

$$Risk\ of\ HIV\ infection = 0.31 + 0.28(45)$$

$$+\ 0.08(65) - 0.05(9)$$

$$-\ 0.11(8) - 0.14(10)$$

$$= 15.39 \quad \text{(ii)}$$

From equations (i) to (ii) it is clear that an increase in the age of the pregnant woman, provided all the other demographic characteristics are kept constant, results in an increase in the risk of HIV infection.

In summary, a regression model can be used for the following purposes,

a. *Interpolation:* where a regression model is used to approximate values of the dependent variable for values of independent variables that are present in the dataset,

b. *Extrapolation*: where the regression model is used to predict values of the dependent variable for values of the independent variable that are outside the range of the dataset under investigation,

c. *Statistical and mathematical relationship between predictor and response variables*: the regression model can be used to determine the extent to which the independent variables influence the dependent variable,

d. *Time series analysis*: in this situation, the regression model can be used to determine how the dependent variable changes over time provided all the independent variables are kept constant.

3.7. Limitations of regression modelling

a. A regression model cannot be used to detect cause and effect between independent and dependent variables,

b. The assumptions underlying the regression modelling exercise are fundamental to the accuracy and trustworthiness of the regression model,

c. It is important and paramount for the experimenter to fully understand the relationship between the different variables prior to building the regression model.

References

1. Sibanda, W., Pillay, V., Danckwerts, M. P., Viljoen, A. M., van Vuuren, S., & Khan, R. A. (2004). Experimental design for the formulation and optimization of novel cross-linked oilispheres developed for in vitro site-specific release of Mentha piperita oil. *AAPS PharmSciTech*, *5*(1), 128-141.

Inspirational Inventions

Thomas Elkin

Thomas Elkins designed a device that helped with the task of preserving perishable foods by way of refrigeration. Elkins' device utilized metal cooling coils which became very cold and would cool down items which they surrounded. The coils were enclosed within a container and perishable items were placed inside. The coils cooled the container to a temperature significantly lower than that inside of a room thereby keeping the perishable items cool and fresh for longer periods of time.

Chapter 4: Screening Design

This chapter introduces the concept of screening designs, steps involved in screening design problem formulation and examples of screening designs.

4.1. What is screening design?

Screening designs are fundamentally used for the identification of main effects. They are used for the identification of a few significant factors from a list of many potential factors. Screening designs constitute the first steps of any experimental process. The factors with the greatest effects on the response variable can then be studied further using advanced and more sensitive experimental designs such as response surface methodologies.

4.1.1. Examples of screening designs
a) Full factorial designs

These are balanced experimental designs exhibiting all the different combinations of the factor levels. It is important to note that in full factorial designs, main and all interaction effects are not confounded or aliased. As a result of the lack of confounding of effects, the resolution of full factorial designs is infinity. The advantage of a full factorial design is that all factor combinations result in the determination of all main and interaction effects.

b) Fractional factorial designs

These are experimental designs that are comprised of a fraction of the total number of experimental runs in a full factorial design. The fraction of experiments is selected to exploit the sparsity-of-effects principle to demonstrate the main effects of a design while utilizing a subset of the full factorial design with regards to experimental runs. A fractional factorial design can be represented as follows;

$$l^{(m-n)} \qquad (i)$$

The advantage of a fractional factorial design is that there is a reduction in the number of experiments to be conducted. The disadvantage of the fractional factorial design is that confounding of effects is increased. Confounding of effects means that effects become mixed with each other and in some situations main effects are confounded with two-factor interactions. The extent to which factors are confounded is termed resolution.

Table 4.1: Different resolutions of experiments and their applications

Resolution	Features	Purpose
1	• These are experiments with only one run. • Unable to distinguish between high and low levels of a factor	Not useful
2	*Main effects* are aliased with *main effects*	Not useful
3	*Main effects* aliased with *two-factor interactions.*	Recommended for robustness testing
4	• *Main effects* not aliased with two-factor interactions • Two-factor interactions aliased with each other	Recommended for screening
5	Main effects not aliased with two-factor interactions Two-factor interactions not aliased with each other	As good as full factorial design

c) *Plackett-Burman designs*

Plackett-Burman designs are experimental designs that were developed by Robin L. Plackett and J.P. Burman in 1946. These designs are used to identify the most important factors affecting an experimental process. Plackett-Burman designs are ideal for the study of main effects provided the two-factor interactions are considered to be negligible.

Therefore Plackett-Burman designs, full factorial and fractional factorial designs are useful for the fitting first-order models which have the ability to demonstrate linear effects. Furthermore, the addition of center-points to the above experimental designs allows the experimental to detect the existence of 2nd order effects.

4.2. DoE Research Problem Formulation

4.2.1. Selection of experimental objective

There are various objectives for an experimental design, such as comparative, screening, response surface and regression model objectives.

A comparative objective entails investigating of a few factors to determine whether a given factor is important in influencing the response variable. On the other hand, screening designs are conducted to identify a few important main factors from the less important factors in an experimental process. Response surface modelling is aimed at evaluating the interaction and quadratic effects of independent experimental factors. The regression model is aimed at developing a statistical and mathematical model of the response with respect to selected predictor variables.

4.2.2. Selection of experimental factors

The primary purpose of a design-of-experiments (DoE) procedure is to estimate the effects of various predictor variables on identified response (dependent) variables.

a) Two-level Experimental designs

These are the most common experimental designs and are ideal for screening purposes. Two-level design-of-experiments are ideal for fitting linear relationships between independent variables and dependent variables.

(i) Two-level full factorial design

The total number of experimental runs for a two-level full factorial design can be determined using the equation below;

$$Number\ of\ experimental\ runs = 2^k \quad (ii)$$

where 2 = number of levels and k = number of factors

    (ii)   Two-level fractional factorial design

The total number of experimental runs for a two-level fractional factorial design can be determined using the equation;

Number of experimental runs for fractional factorial design = Fraction $* 2^k$

Half fractional factorial design $= \dfrac{1}{2}*2^k = (2^{-1} * 2^k) = 2^{(k-1)}$ *(iii)*

Quarter fractional factorial design $= \dfrac{1}{4}*2^k = (4^{-1} * 2^k) = (2^2 * 2^k)$

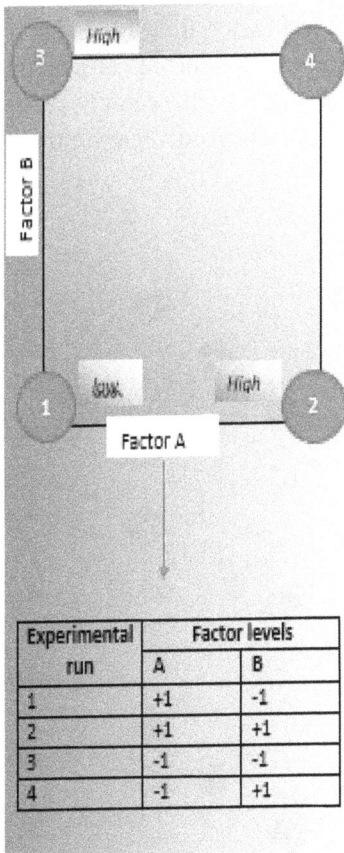

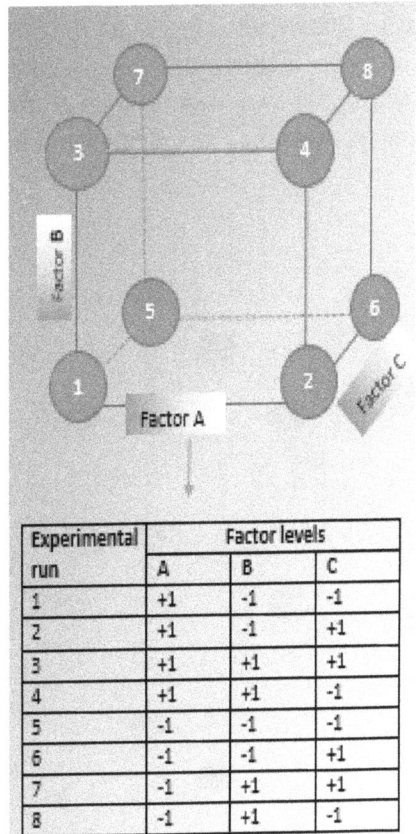

Experimental	Factor levels	
run	A	B
1	+1	-1
2	+1	+1
3	-1	-1
4	-1	+1

Experimental	Factor levels		
run	A	B	C
1	+1	-1	-1
2	+1	-1	+1
3	+1	+1	+1
4	+1	+1	-1
5	-1	-1	-1
6	-1	-1	+1
7	-1	+1	+1
8	-1	+1	-1

(a)   (b)

$$= \ 2^{(k\,+\,2)} \ \textit{(iv)}$$

Fig. 4.1: Schematic representations of the two-level and three-level full factorial designs respectively

(b) *Three-level full factorial design*

Three-level factorial designs have the ability to fit a curved response function such as a quadratic function. The total number of experimental runs in three level factorial design can be investigated as follows;

Number of experimental runs $= 3^k$

Where $3 = number\ of\ levels\ and\ k = number\ of\ factors$

The three levels of the experimental design are the low, center and high levels. The disadvantage with the three-level full factorial design is that there is a large number of experimental runs to be conducted. Research has however, shown that two-level factorial designs with center-points are equally capable of demonstrating the presence of curvature with

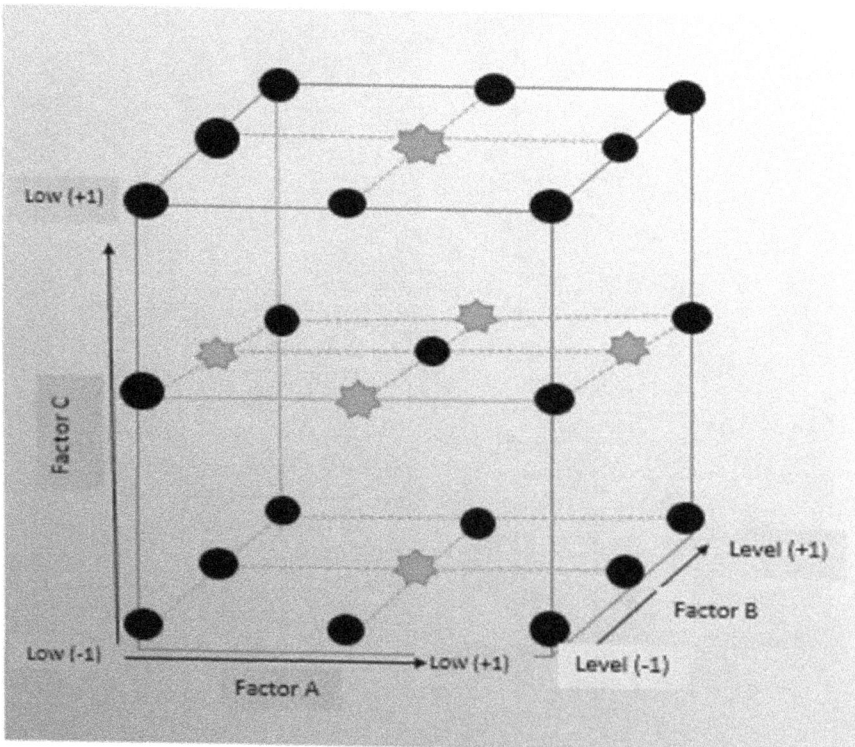

comparatively fewer experimental runs compared to a three-level factorial design.

Fig. 4.2: Schematic representation of a 3^k full factorial design

4.2.3. Selection of an experimental response

It is important for an experimenter to select the most appropriate responses in accordance with experimental goals.

Demonstration of the Selection of an Experimental Response

In a study, a two-level fractional factorial design was employed to investigate the effect of demographic characteristics on the risk of HIV infection amongst pregnant women attending antenatal clinics in South Africa.

In the above study, the risk of HIV infection was selected as the most appropriate response and was found to be highly sensitive to changes in the age of the pregnant women and her level of education.

The selection of the regression model depends on the purpose of the DoE. In this instance, the purpose of the study was to screen out the most important factors affecting the risk of HIV infection amongst pregnant women attending antenatal clinics in South Africa.

An appropriate regression model for a fractional factorial design is a linear model with either main effects only or linear model with both main effects and interaction effects. In this example, a linear model with only main effects model was selected, as shown in the equation below;

$Risk\ of\ HIV\ infection = 0.31 + 0.28W + 0.08F + 0.05G - 0.11P - 0.14E$

Where; $W = woman's\ age,\ M = Male\ partner's\ age,$

$G = Gravidity,\quad P = Parity, E = Education\ and\ S = Syphilis$

Example of a Screening Design

*In **Case-study 1**, where an experiment using a Plackett-Burman design to develop a formulation of crosslinked calcium-aluminium-alginate-pectinate gelispheres for in-vitro site-specific release of Mentha piperita, an essential oil for the treatment of irritable bowel syndrome, numerous physicochemical and textural properties were tested as responses(**Sibanda et al.2004**)*

a) **Design matrix:** The Plackett-Burman experimental matrix illustrating the various formulations tested was developed as shown in Table 2 below. The formulations indicate the reaction inputs and their concentrations. The aim was to select the formulation that produced the best in-vitro drug release profile.

Table 4.2: Plackett-Burman design matrix with 5 predictor variables and 14 experimental runs

Oilisphere Formulation	Sodium alginate (% wt/vol)	Pectin % wt/vol	Calcium chloride (% wt/vol)	Aluminum chloride (%wt/vol)	Aluminum sulfate (%wt/vol)	Crosslinking Reaction Time (Hours)
F1	0	1.5	4	4	0	6
F2	1.5	0	4	0	0	0.5
F3	1.5	0	0	0	4	6
F4	0	1.5	0	0	0	6
F5	1.5	1.5	0	4	0	0.5
F6	1.5	1.5	0	4	4	0.5
F7	0	0	0	0	0	0.5
F8	1.5	0	4	4	0	6
F9	0	0	4	4	4	0.5

F10	0.75	0.75	2	2	2	3.25
F11	0.75	0.75	2	2	2	3.25
F12	0	0	0	4	4	6
F13	1.5	1.5	4	0	4	6
F14	0	1.5	4	0	4	0.5

b) **Experimental Responses:** *Four response variables were identified and measured namely, total fracture energy of the crosslinked gelisphere, gelisphere matrix hardness, hydrated and unhydrated gelisphere matrix resilience, as shown in Table4.2. Total fracture energy was found to be in close agreement with the predicted values of the optimized formulation.*

c) **Regression analysis:** *R^2 and Durbin-Watson statistics were used as regression diagnostics to determine the validity of the model.*

i. R^2 is a measure of the closeness of the fitted regression model to the observed data.

ii. Durbin-Watson (D) represents a test for autocorrelation in residuals from a statistical regression analysis. D is measured between 0 and 4, where 2 indicates lack of autocorrelation, O indicates positive autocorrelation and 4 depicts negative autocorrelation.

Table 4.3: Response variables

Experimental responses	R^2	Durbin-Watson (D)
a. Unhydrated matrix resilience	0.74	1.45
b. Hydrated matrix resilience (%)		
pH 3 (%)	0.63	1.60
pH 6.8 (%)	0.41	1.94
c. Matrix-hardness (N/m)	0.74	2.20

d. Total-Fracture Energy (J)	0.88	1.75

d) Plot of Experimental vs Predicted Responses

The highest correlation between experimental and predicted data response values was observed for total fracture energy compared to the other responses variables generated by the Plackett-Burman design· As a result,

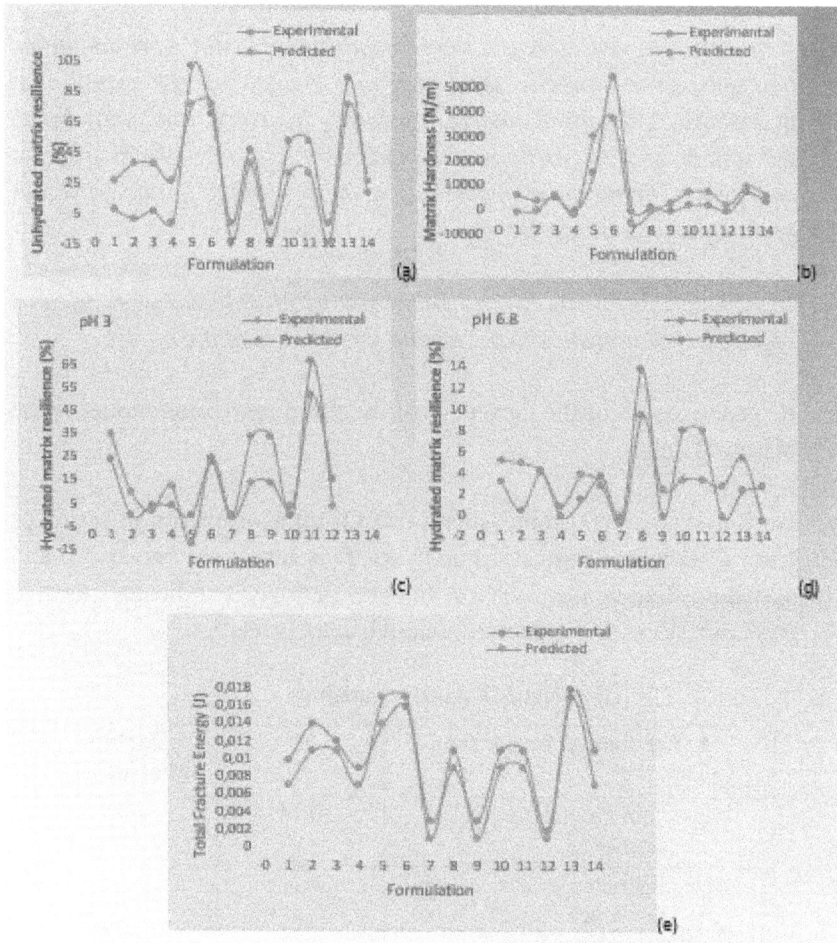

total fracture energy was selected for the optimization objective.

Fig. 4.3: Correlation between experimental responses and responses predicted by the Plackett-Burman model for a) unhydrated matrix resilience, b) matrix hardness, c) hydrated matrix resilience in buffer medium pH 3, and d0 hydrated matrix resilience in buffer medium pH 6.8 and e) total fracture energy

On the basis of regression analysis, measured total fracture energy response was selected for optimization. The regression model developed for total fracture energy was found to be;

$$TFA = 0.08 + 0.0006 NaAlg$$
$$+ 0.011P + 0.081 CaCl_2$$
$$+ 0.598 AlCl_3$$
$$+ 0.367 Al_2(SO_4)_3$$
$$+ 0.62 CRT$$

where:

$TFA = total\ fracture\ energy,\ NaAlg = sodium\ alginate,$

$CaCl_2 = calcium\ chloride$

$AlCl_3 = aluminium\ chloride,\ Al_2(SO_4)_3 = aluminium\ sulphate$

$CRT = crosslinking\ reaction\ time$

Optimization of the selected response (total fracture energy) was accomplished using a Lagrangian method. The composition of the optimized gelispheres is shown below;

Table 4.4: Optimized formulation

Formulation	Concentration (%wt/vol)
Sodium alginate	1.5
Pectin	1.5
Calcium chloride	4
Aluminum chloride	4
Aluminum sulphate	4
Crosslinking reaction time (hrs)	3
Optimized Formulation	
	Total fracture Energy (J)
Experimental	0.0115 ± 0.003

Predicted	0.0107

The above table shows that constrained optimization developed an improved gelisphere formulation indicated by the excellent correlation between predicted and experimental total fracture energy response variable. Therefore optimization studies using total matrix fracture energy enabled the formulation of a pH-dependent zero-order releasing oilisphere system for in vitro site-specific release of Mentha piperita for the treatment of irritable bowel syndrome.

SOLVED PROBLEMS

Problem 1: There is an increase in the use of nanotechnology in medicine and more specifically drug delivery. Pharmaceutical companies are using nanoparticles to reduce toxicity and side effects of drugs. Research has also shown that nanoencapsulation of drugs increases their efficacy, specificity and targeting ability.

A pharmaceutical company sought to use a half fractional factorial design to determine how the amount of drug released by a nanoparticle after 30 minutes was affected by the following nanoparticle physicomechanical characteristics; size of nanoparticle, dissolution at pH 6.8, drug encapsulation at pH 6.8, fracture force and hardness. The levels of the five factors are shown in Table below.

Table 4.5: Physicomechanical properties of nanoparticles

Factor	Units	Actual Levels		Coded Levels		Response
Size	nm	2.5	3.0	-1	+1	DrugReleased
Dissolution pH		3.0	3.5	-1	+1	(30 mins).
Drug Encapsulation	%	70	80	-1	+1	
Fracture Energy	Joules	0.006	0.008	-1	+1	
Matrix Hardness	N/m	10 000	20 000	-1	+1	

The total number of experimental runs (N) for a full factorial design with five (5) factors, each measured at 2 levels is $N = 2^k = 2^5 = 32$. Therefore, a half-fractional factorial will have $2^{k-1} = 2^{5-1} = 16$ experimental runs, as shown in the design matrix below.

Size	Dissolution rate	Drug Encapsulation	Fracture Energy	Matrix Hardness	Drug Release
-1	-1	-1	-1	1	0.65
1	-1	-1	-1	-1	0.81
-1	1	-1	-1	-1	0.74
1	1	-1	-1	1	0.23
-1	-1	1	-1	-1	0.32
1	-1	1	-1	1	0.74
-1	1	1	-1	1	0.725
1	1	1	-1	-1	0.53
-1	-1	-1	1	-1	0.86
1	-1	-1	1	1	0.65
-1	1	-1	1	1	0.93
1	1	-1	1	-1	0.47
-1	-1	1	1	1	0.306
1	-1	1	1	-1	0.445
-1	1	1	1	-1	0.35
1	1	1	1	1	0.23

Table 4.6: A half-fractional factorial design matrix

Show:

a) Main Effects

b) Interaction plots

c) Pareto plots

d) Surface Response plot

e) Contour plot

a) Main effect

The main effects plot shows the response mean for each physicomechanical factor level connected by a line. A horizontal line indicates that there is no main effect. A line that is not horizontal, shows the presence of a main effect.

The steeper the slope of the line, the greater the main effect. However, main effects are not capable of indicating the presence of factor interactions. From this example, nanoparticle size, dissolution rate, drug encapsulation and fracture energy affect drug release, because the lines are not horizontal.

However, for the four physicomechanical properties (nanoparticle size, dissolution rate, drug encapsulation and fracture energy), the low levels (-1) have a higher effect on drug release that the higher levels (+1). Matrix hardness has little to no effect on drug release based on a near horizontal line.

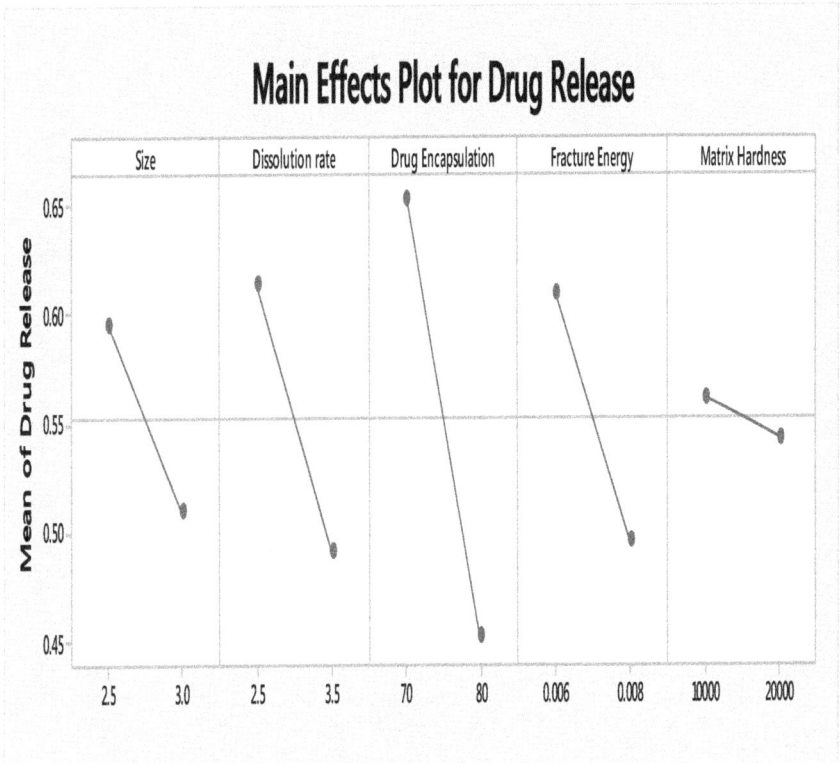

Fig. 4.4: Main effects affecting drug release

b) Interaction Plot

The interaction plots is used to indicate the relationship between physicomechanical variables and the continuous response (Drug release). Interaction plot evaluates how the interactions affect the relationship between the physicomechanical factors and the response. Parallel lines indicate that there is no interaction, while nonparallel lines show that an interaction occurs. Therefore, the more nonparallel the lines are, the greater the strength of the interaction. In Fig. 4.5, significant interaction is observed between size*dissolution, size*fracture energy, size*matrix hardness. However, no interaction was observed between dissolution*fracture energy.

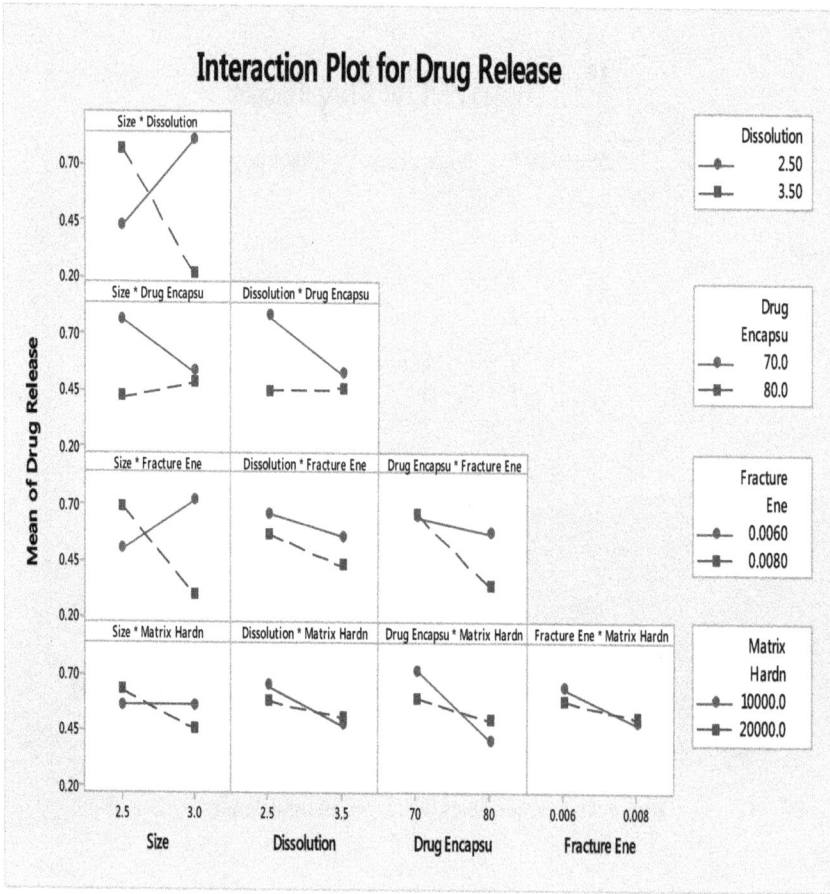

Fig. 4.5: Interaction of physico-mechanical factors affecting drug release

c) Pareto Chart

The purpose of the Pareto chart is to highlight the most important among the five (5) physico-mechanical properties of the nanoparticles and their interactions. The Pareto chart assists us in visualizing which factors comprise the 20% that are most critical, i.e. in this study, the interaction of size*dissolution. In general, the Pareto principle states that 80% of the effects on a given response are caused by 20% of the investigated factors.

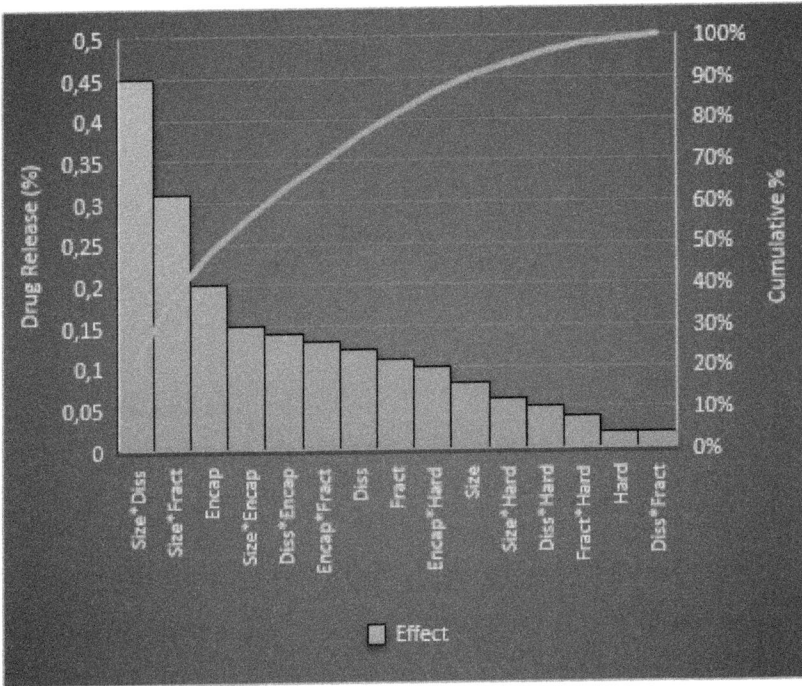

Fig. 4.6: Pareto effect showing the contribution of each factor to drug release

d) **Surface Response plot**

A surface plot provides a three-dimensional view of the response (drug release) against two of the physicomechanical properties of the nanoparticles. A surface response plot has curvature as a result of the quadratic terms in the model. In the plot below, the highest values of drug release corresponding to high values of dissolution pH and low values of nanoparticle size.

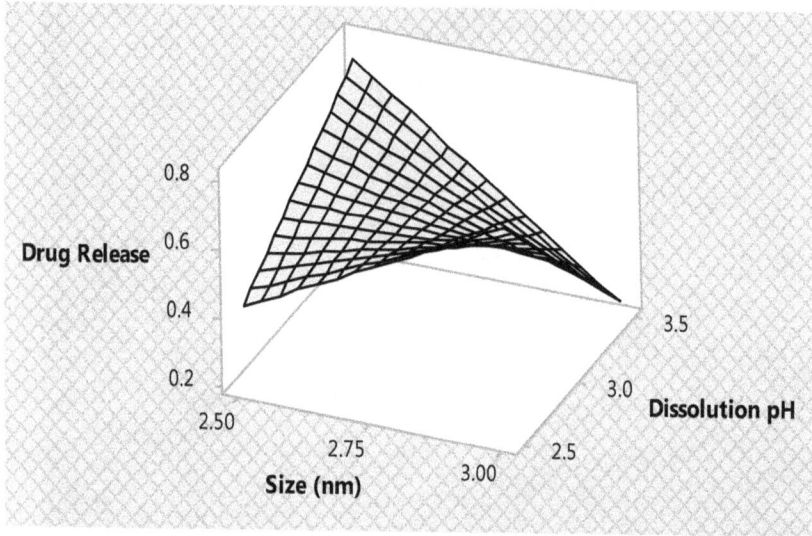

Fig. 4.7: Response surface plot of drug release (size vs dissolution)

In this plot, the high drug release is achieved at low fracture energy and low matrix hardness.

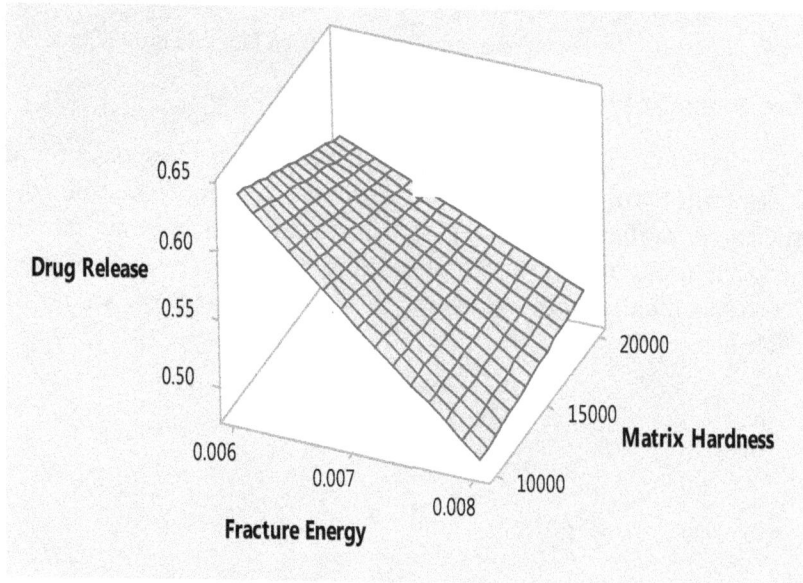

Fig. 4.8: Response surface plot of drug release (drug release vs fracture energy)

e) Contour plot

Contour plots are vital tools that provide a two-dimensional view where all points with equal response are linked to produce contour lines of constant responses. Therefore, a contour plot provides valuable information on the desired response values and operating conditions. From the plot below, it's clear that a decrease in dissolution and decrease in matrix hardness results in a decrease in an increase in drug release.

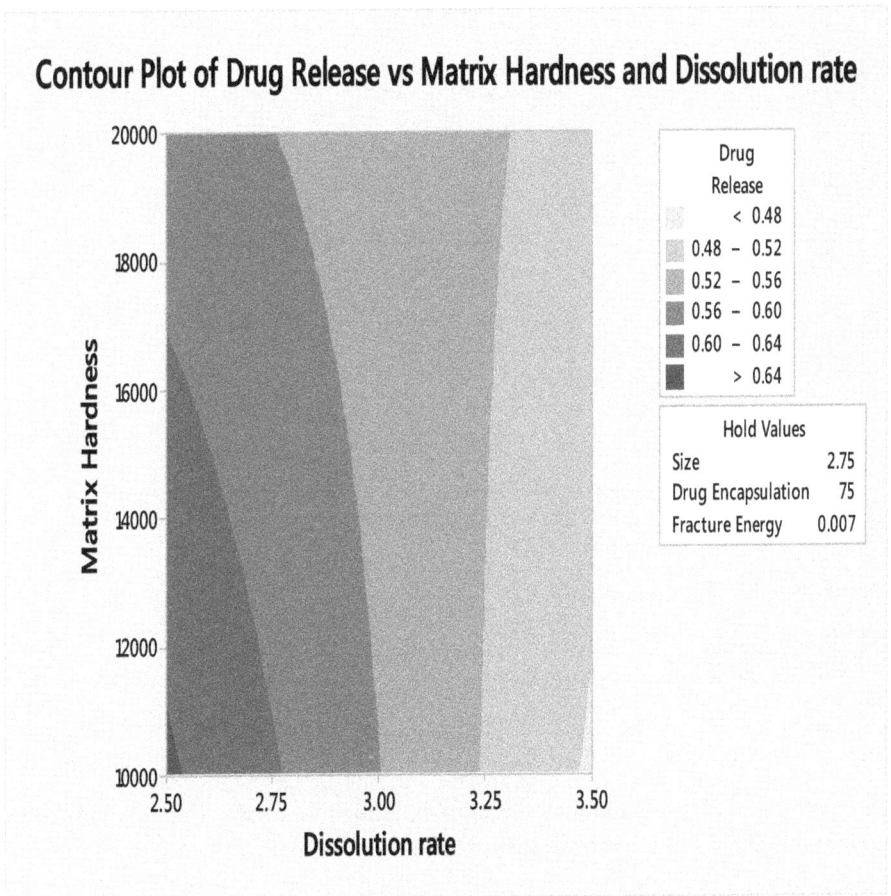

Fig. 4.9: Contour plot of drug release drug release vs matrix hardness and dissolution rate

Problem 2: A health and beauty company sought to determine the degradation of its face and body lotion at a temperature of 70°C as a measure of its stability. The face and body lotion comprised of the following ingredients.

Table 4.7: Ingredients of face and body lotion

Ingredient	Low Level		High Level	
	Quantity (%)	Coded	Quantity (%)	Coded
Methylglucose stearate	2.0	-1	4.0	+1
Linoleic acid	3.0	-1	6.0	+1
Glyceryl stearate	2.1	-1	4.2	+1
Tocopherol	0.5	-1	1.0	+1
Citric acid	0.25	-1	0.5	+1
Xanthan Gum	0.05	-1	0.10	+1

A fractional factorial design resolution 3, with 6 factors at 2 levels was used to develop an experimental design matrix. A fractional factorial design resolution 3 is a screening design, often used to identify important experimental factors from many potential factors influencing a response. These designs estimate main effects only, and assume interactions between factors are not significant. The number of experiments in a two-level fractional factorial design resolution 3, is determined by 2^{n-3} where n = 6. Therefore, the total number of experiment runs is 8.

Design matrix

Table 4.8: A fractional factorial design matrix

Exp. No.	Methylglucose stearate	Linoleic acid	Glyceryl stearate	Tocopherol	Citric acid	Xantham Gum	Degradation rate
1	+1	-1	-1	+1	-1	+1	0.1
2	+1	+1	-1	-1	+1	-1	0.3
3	+1	+1	+1	-1	-1	+1	0.2
4	-1	+1	+1	+1	-1	-1	0.5
5	+1	-1	+1	+1	+1	-1	0.3
6	-1	+1	-1	+1	+1	+1	0.2
7	-1	-1	+1	-1	+1	+1	0.27
8	-1	-1	-1	-1	-1	-1	0.4

Analysis of Experimental data

a) Histogram

The shape of the response (degradation rate) with a slight skew to the right, indicates that transformation is required. Based on the shape of the histogram plot, it was decided the data using a log-transformation, as shown in Fig. 4.10

Experiment No.	Degradation Rate (x_i)	Log Transformed Degradation Rate (Log x_i)
1	0.10	-2.30
2	0.30	-1.20
3	0.20	-1.61
4	0.50	-0.69
5	0.30	-1.20

6	0.20	-1.61
7	0.27	-1.31
8	0.40	-0.92

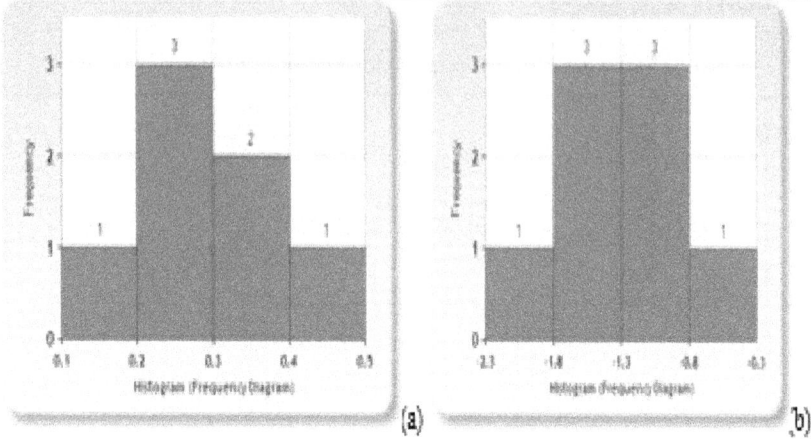

(a) (b)

Fig. 4.10: Histogram plot of (a) degradation rate data, (b) Log transformed degradation data

b) Box-Whisker plot

Box-whisker plot confirms the histogram plot findings that the response has a slight skew to the right and requires transformation

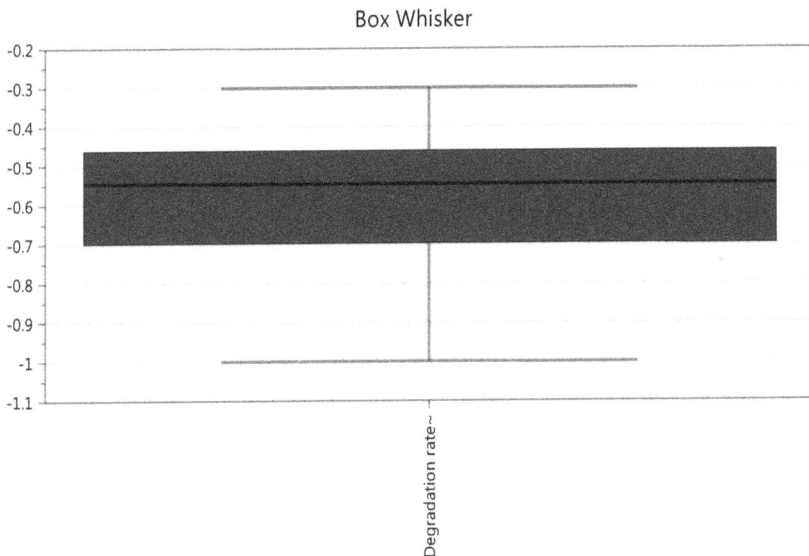

Fig. 4.11: Box-whisker plot

Fitting the model to the data-ANOVA

ANOVA is concerned with estimating different types of variability in the response variable and then comparing such estimates with each other by means of F-test. The ANOVA shows that the regression model is statistically significant ($P<0.05$). All the predictor variables are statistically significant except linoleic acid and tocopherol ($P>0.05$). R^2 and R^2 adjusted values were 0.995 and 0.98 respectively, at 95% confidence level.

Table 4.9: ANOVA table

Degradation rate	DF	SS	MS	F	P
Total	8	3.09624	0.387031		
Constant	1	2.77455	2.77455		
Total corrected	7	0.321697	0.0459567		
Regression	5	0.320122	0.0640245	81.3168	0.012
Residual	2	0.0015747	0.000787		

115

Coefficient plot

A regression coefficient gives information on the size and direction of the relationship between independent and dependent (response) variables. In other words, coefficients are an indication of the value used to multiply the regression terms in a regression model. A coefficient is therefore an average change in the term, provided all the other terms are held constant. The sign of the coefficient indicates the direction of the relationship between predictor variables and response variables. However, the size of the coefficient does not indicate whether a predictor variable is statistically. Therefore, to determine statistical significance of the predictor variables, P values should be used. Coefficient plots are graphical representations of the model terms. Coefficient plots are used to determine the significance of the model terms. Significant model terms have large values and the error bars do not cross the y=0 line. Xanthine has the greatest, but negative effect on the rate of degradation and the coefficient is statistically significant. In other words, an increase in Xanthine results in a decrease in the rate of degradation.

Table 4.10: Coefficient table

Degradation rate	Coefficient (Scaled and centralized)	P-value
Constant	-0.588913	0.000283654
Methylglucose stearate	-0.0972689	0.0102427
Linoleic acid	0.0334507	**0.0778264**
Glyceryl stearate	0.0660342	0.0218338
Tocopherol	-0.0418067	0.05196
Xantham Gum	-0.152731	0.00419259

116

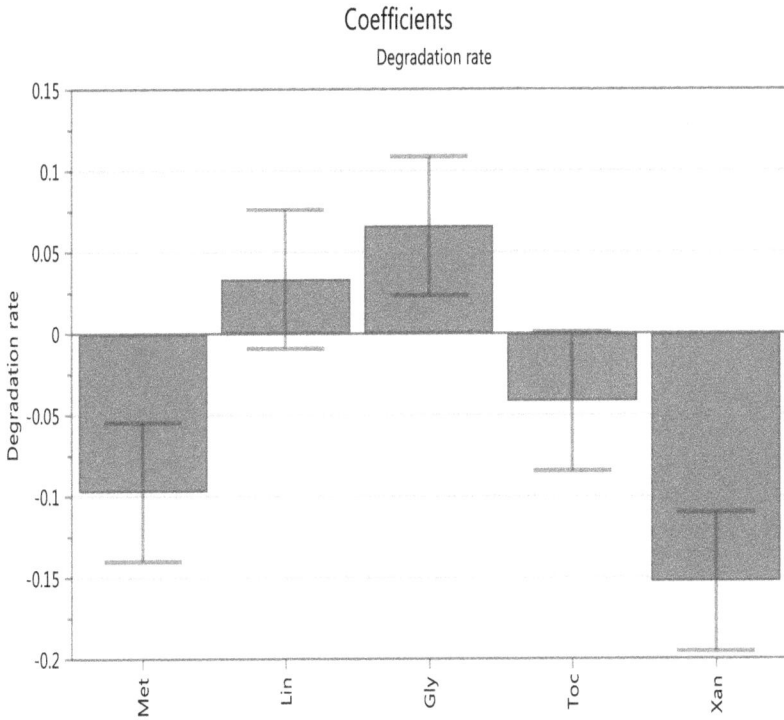

Fig. 4.12: Coefficient plot

Pareto chart

As was explained in Solved Problem 1, the purpose of the Pareto chart is to show the most important variables among a large set of variables. According to the Pareto chart (Fig. 4.13), the most important factor influencing degradation rate is Xantham gum followed by methylglucose. The least influential factor is tocopherol.

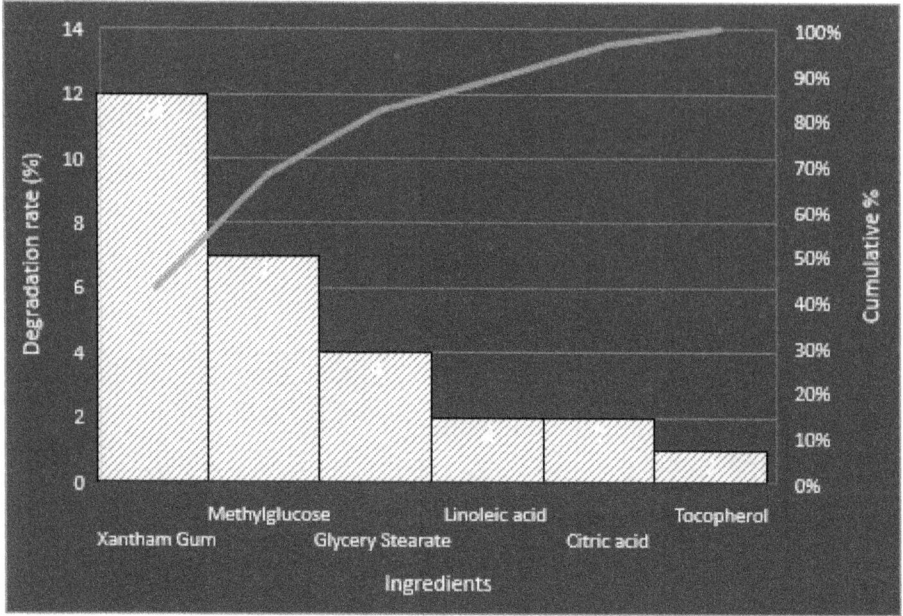

Fig. 4.13: Pareto plot

Main effect plots

A main effect is the effect of a single independent variable on a dependent variable, ignoring all other independent variables. Xantham gum has the greatest main effect of 0.19-0.36=-0.17. This means that as the amount of Xantham gum changes from low (-1) to high (+1), there is a corresponding decrease in the degradation rate of the face and body lotion by 0.17. Only glycerol stearate and linoleic acid have positive main effects of 0.07 and 0.03 respectively. However, it is not advisable to interpret main effects without considering the interaction effects.

118

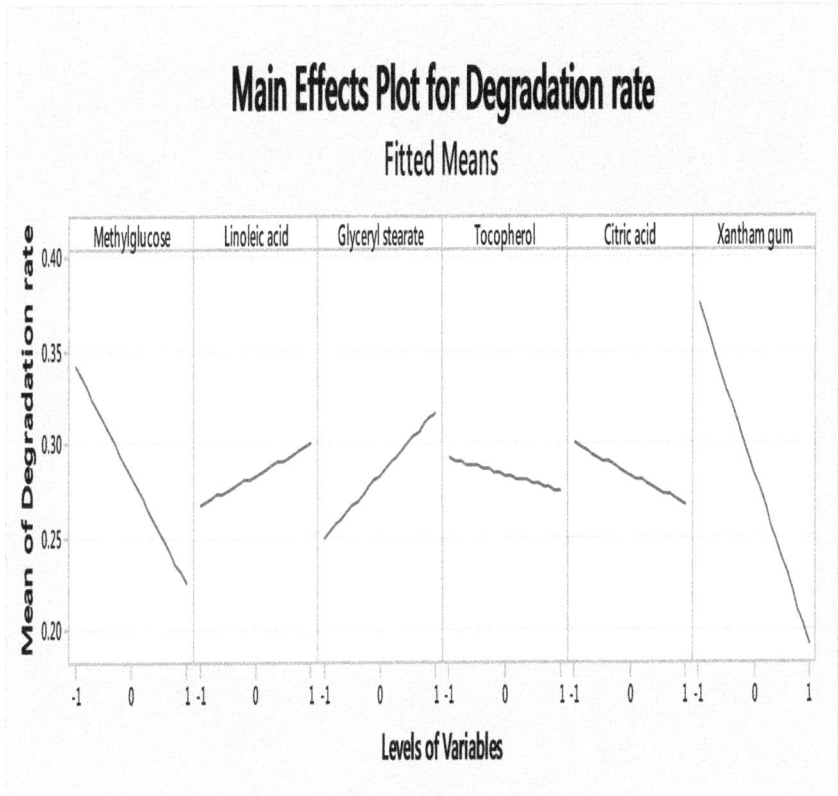

Fig. 4.14: Main effects plot

Interaction plot

An interaction plot is used to illustrate that the relationship between one categorical variable predictor variable and a continuous response, is dependent on another categorical variable. In general, parallel lines indicate that there is no interaction (e.g. methylglucose*glycerol stearate), while non-parallel lines indicate there is interaction (e.g. tocopherol*citric acid). The more non-parallel, the lines are, the greater the strength of the interaction (e.g. tocopherol*xanthum gum and citric acid*xanthum gum).

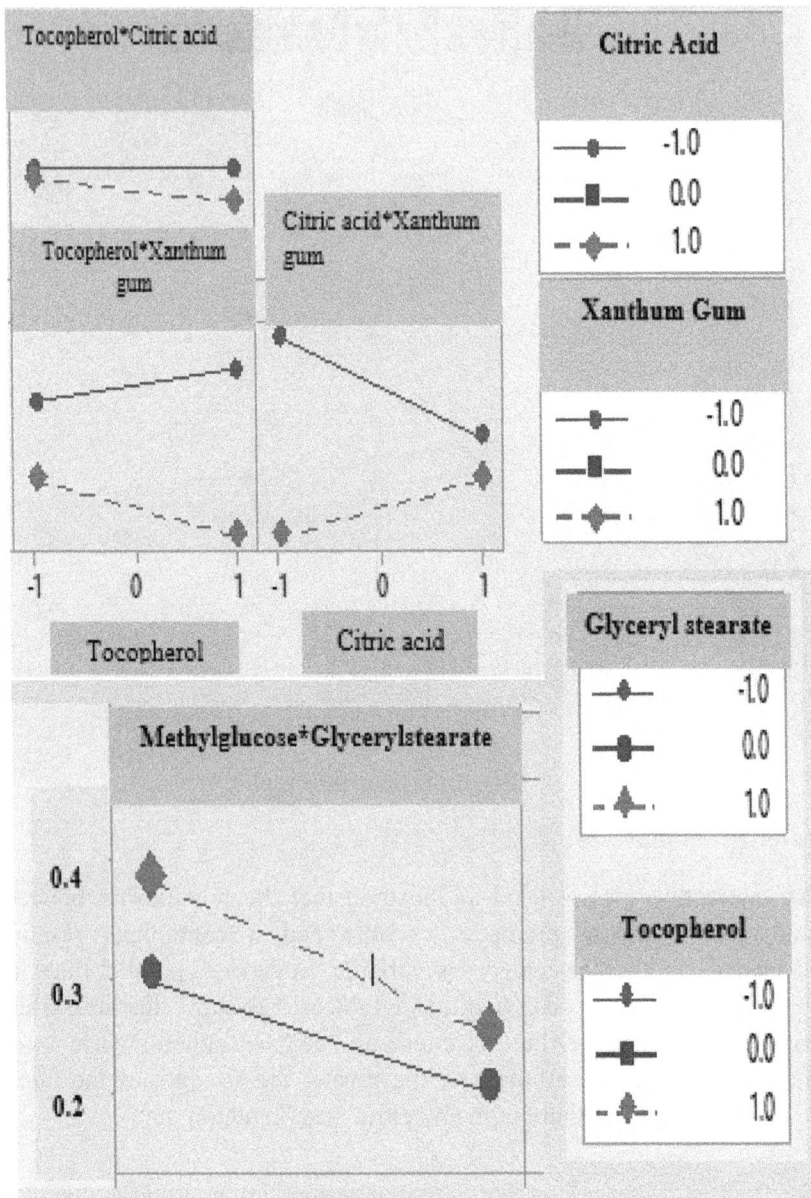

Fig. 4.15: Interaction plots

Contour plots

The contour plot shows that low levels (-1) of methyl glucose stearate are associated with increased degradation rate of the face and lotion formulation. The highest degradation at low level (-1) of methyl glucose stearate occurs at the high (+) level glyceryl stearate and at low (-1) of tocopherol.

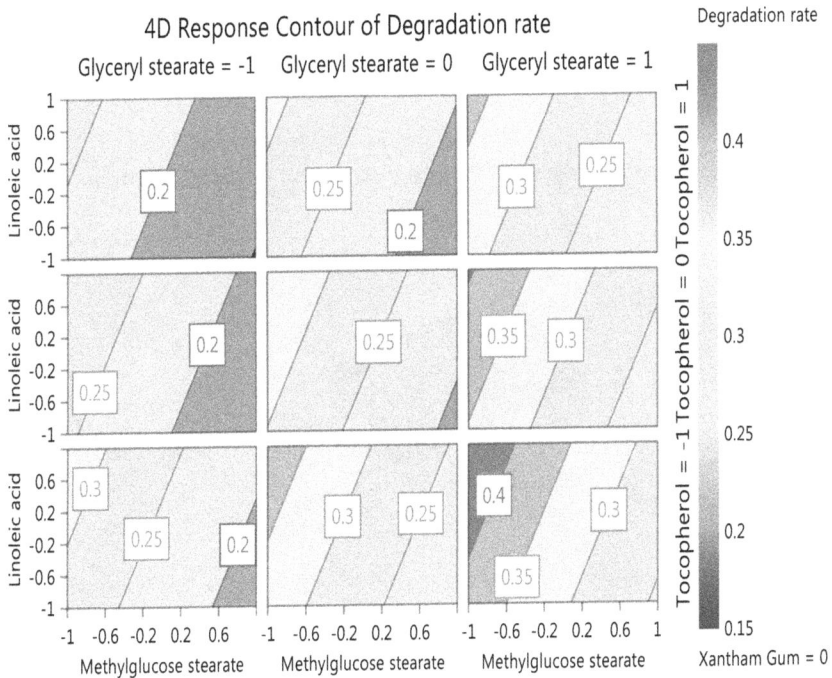

Fig. 4.16: 4 D contour plots

Response surface plot

A surface plot displays the three dimensional relationship in two dimensions with independent variables in x and y axes and the response variable in z axis, as shown in Fig. 4.17.

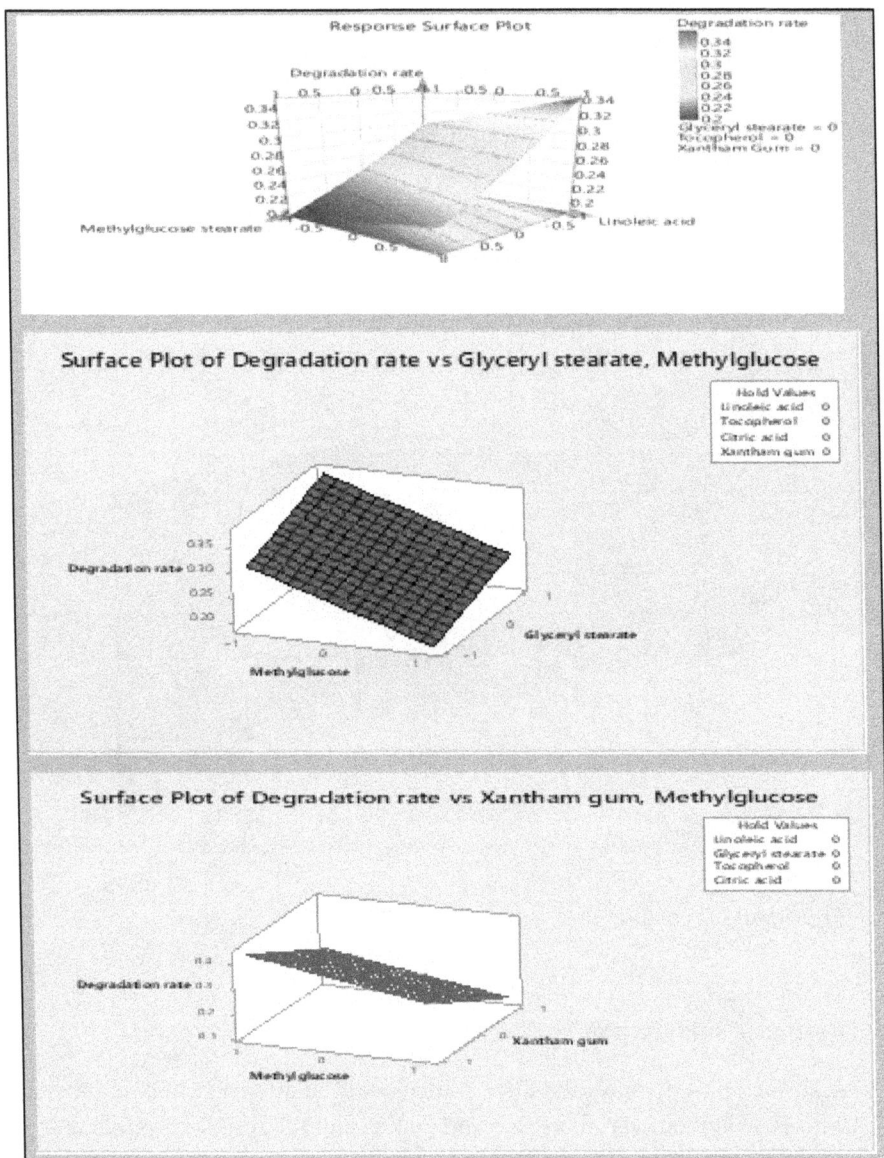

Fig. 4.17: Response surface plots

122

Problem 3: A study was conducted to determine the efficacy (y_i) of a mosquito insecticide using four different species of malaria transmitting mosquitoes, a_1, a_2, a_3, a_4. Insecticide efficacy was measured by the reduction in the population of mosquitoes. The researchers decided to use a Plackett-Burman design to create the design matrix and it was decided to have eight (8) experiments.

Calculation by Hand

Based on a Plackett-Berman design, an experiment with $n = 8$ runs can only have $n - 1 = 7$ factors. However, in this research there are only four (4) factors, namely a_1, a_2, a_3, a_4. Now since, there are only four variables, the other three (3) factors were therefore dummy factors, d_1, d_2, d_3, d_4. Dummy variables were used to determine random errors. The Plackett-Burman utilized two levels for each factor, with the higher level (+1) and lower level (-1).

a) **Design matrix**

Table 4.11: Plackett-Burman design matrix

Experiment	a_2	d_1,	a_2	d_2	a_3,	d_3	a_4,	Response (Efficacy mosquito insecticide)
1	+1	-1	-1	+1	-1	+1	+1	10
2	+1	+1	-1	-1	+1	-1	-1	9
3	+1	+1	+1	-1	-1	+1	-1	10
4	-1	+1	+1	+1	-1	-1	+1	9
5	+1	-1	-1	+1	+1	-1	-1	8
6	-1	+1	-1	+1	+1	+1	-1	7
7	-1	-1	+1	-1	+1	+1	+1	7

8	-1	-1	-1	+1	-1	-1	+1	7

b) Calculation of EFFECT of each factor by Hand:

$$\frac{2[\Sigma(y^+) - \Sigma(y^-)]}{N}$$

Where $N = total\ number\ of\ experiments$

$y^+ = responses\ when\ factor\ is\ at\ its\ high\ level\ (+1)$

$y^- = responses\ when\ factor\ is\ at\ its\ high\ level\ (-1)$

e.g. (i) Effect size for factor $a_2 =$

$$\frac{2[\Sigma(y^+) - \Sigma(y^-)]}{N}$$

$$\frac{2[(10 + 9 + 10 + 8) - (9 + 7 + 7 + 7)]}{8}$$

1.75

N.B. Using the same formula effect sizes of the other **factors** and **dummy variables** were calculated as shown in **Table 4.12**. A negative factor effect indicates that a change of factor from high to low value increases the response variable (efficacy of the insecticide) rather than decreasing it.

c) Calculation of Sum of Squares (SS):

$$\frac{N \times (estimated\ effect)^2}{4}$$

e.g. sum of squares of a_2

$$\frac{8 \times (1.75)^2}{4} = 6.125$$

The **sum of squares for all the main effects** $a_1, a_2, a_3, a_4,$ were 6.125, 0.125, 3.125 and 1.125 respectively. Each sum of squares has one (1) degree of freedom (df), meaning that the mean square (MS) values (i.e. variances) are the same as the sum of squares (SS).

The **sum of squares for the dummy variables** d_1, d_2, d_3, d_4, were found to be 1.125, 0.125 and 0.125 respectively. There are three degrees of freedom for the three dummy variables. The mean sum of squares for the dummy variables was determined as follows:

$$MS = \frac{1.125 + 0.125 + 0.125}{3} = 0.458$$

Table 4.12: Plackett-Burman design matrix with Effect sizes and Mean Squares (MS) of Main effects and dummy variables

Experiment	a_2	$d_1,$	a_2	d_2	$a_3,$	d_3	$a_4,$	Response (**M**osquito insecticide Efficacy)
1	+1	-1	-1	+1	-1	+1	+1	10
2	+1	+1	-1	-1	+1	-1	-1	9
3	+1	+1	+1	-1	-1	+1	-1	10
4	-1	+1	+1	+1	-1	-1	+1	9
5	+1	-1	-1	+1	+1	-1	-1	8
6	-1	+1	-1	+1	+1	+1	-1	7
7	-1	-1	+1	-1	+1	+1	+1	7
8	-1	-1	-1	+1	-1	-1	+1	7
Effect	1.75	0.75	0.25	0.25	-1.25	0.25	0.75	
Mean SS	6.125	1.23	0.125	0.125	3.125	0.125	1.125	

d) Calculation of F-values for each factor using the formula:

$$F\ value = \frac{MSE\ (Variable)}{MSE\ (Dummy\ variables) =} \quad \frac{MSE\ (Variable)}{0.458}$$

If the calculated F-value is higher than the critical F-value from the F-table, the null hypothesis is rejected meaning that the factor under analysis has a significant effect of the response variable.

Table 4.13: ANOVA

Source of variation	df	Sum of squares	Mean squares	F-ratio	Critical values F-	P-value
Model						
a_1	1	6.125	6.125	$\frac{6.125}{0.458} = 13.37$	10.13	Significant
a_2	1	0.125	0.125	$\frac{0.125}{0.458} = 0.273$	10.13	Not significant
a_3	1	3.125	3.125	$\frac{3.125}{0.458} = 6.823$	10.13	Not significant
a_4	1	1.125	1.125	$\frac{1.125}{0.458} = 2.456$	10.13	Not significant
Residual	3	1.374	0.458			

Total						

Problem 4: A Plackett-Burman design was used to study the effect of four factors (A,B,C and D) on the response variable. Each factor was studied at two levels. A total of 12 experiments were developed as shown below:

a)   Plackett-Burman design matrix

A Plackett-Burman design matrix was developed as shown in Table 4.12, with 4 factors, at two levels each and one response variable. A design matrix provides a representation of an experiment that shows factor level combinations and associated response. Each row is a design matrix depicts a treatment combination and each experiment has numerous treatment combinations. As discussed in earlier sections in this Handbook, Plackett-Burman designs were developed by English statisticians Drs. Plackett and Burman in 1946. The assumption in Plackett-Burman designs is that all interactions are insignificant relative to main effects.

Table 4.14: A Plackett-Burman design with factors

Experiment	A	B	C	D	Response
1	-1	1	-1	-1	75.67
2	1	-1	1	-1	102.40
3	-1	-1	1	1	113.71
4	-1	-1	-1	1	72.95
5	1	-1	1	1	102.32
6	1	1	-1	1	113.85
7	-1	1	1	-1	106.51
8	1	-1	-1	-1	72.66
9	-1	1	1	1	68.62
10	-1	-1	-1	-1	75.54
11	1	1	1	-1	109.28
12	1	1	-1	1	75.69

b) Coefficient Plot

A stepwise regression technique was used to fit a full model to the data. As shown in Table 4.15, none of the variables were found to be statistically significant, at p=0.05. The coefficients describe the mathematical relationship between each independent variable and the dependent variable. The p-value indicates whether the relationships are statistically significant.

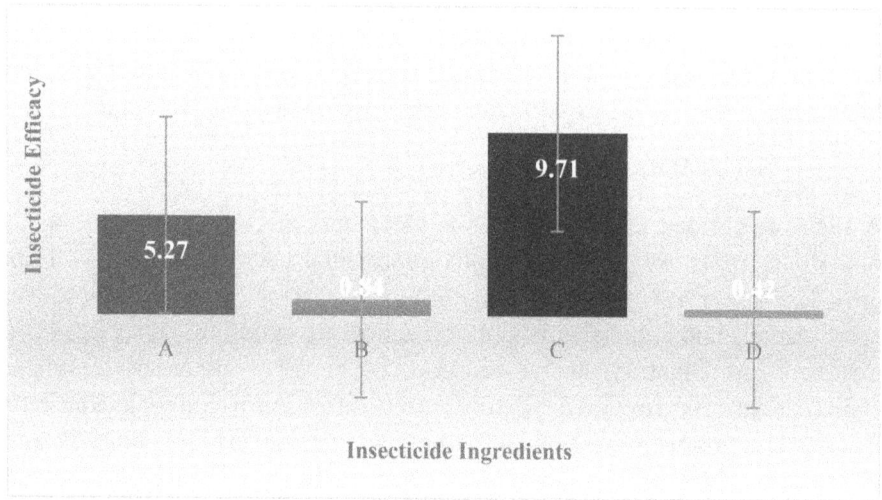

Fig. 4.18: A Plackett-Burman coefficient plot

Table 4.15: A Coefficient table

	Coefficients	Standard Error	t Stat	P-value	Lower 95%	Upper 95%
Intercept	90.77	5.20	17.5	0.00	78.5	103
A	5.27	5.20	1.01	0.34	-7.0	18
B	0.84	5.20	0.16	0.88	-11.5	13

C	9.71	5.20	1.87	0.10	-2.6	22
D	0.42	5.20	0.08	0.94	-11.9	13

Regression Equation

Regression analysis is a technique for studying the relationship between dependent and independent variables. Regression enables the determination of the effect of each independent variable on the dependent variable, controlling for the effect of all other independent variables.

In addition, regression allows for the prediction of the value of dependent variable for a given value of the independent variable. The regression model relating the independent variable to the response variable is shown below,

$$\text{Response} = 90.77 + 5.27\ A + 0.84\ B + 9.71\ C + 0.42\ D$$

Table 4.16: Analysis of Variance (ANOVA)

Source	DF	Adj SS	Adj MS	F-Value	P-Value
Model	4	1474.04	368.51	1.14	0.412
Linear	4	1474.04	368.51	1.14	0.412
A	1	332.85	332.85	1.03	0.344
B	1	8.40	8.40	0.03	0.877
C	1	1130.63	1130.63	3.49	0.104
D	1	2.15	2.15	0.01	0.937
Error	7	2267.44	323.92		

Lack-of-Fit	6	1539.34	256.56	0.35	0.857
Pure Error	1	728.09	728.09		
Total	11	3741.47			

Pareto chart

Factors C has the greatest effect on the response variable followed by Factor A. Factor D, has the least effect.

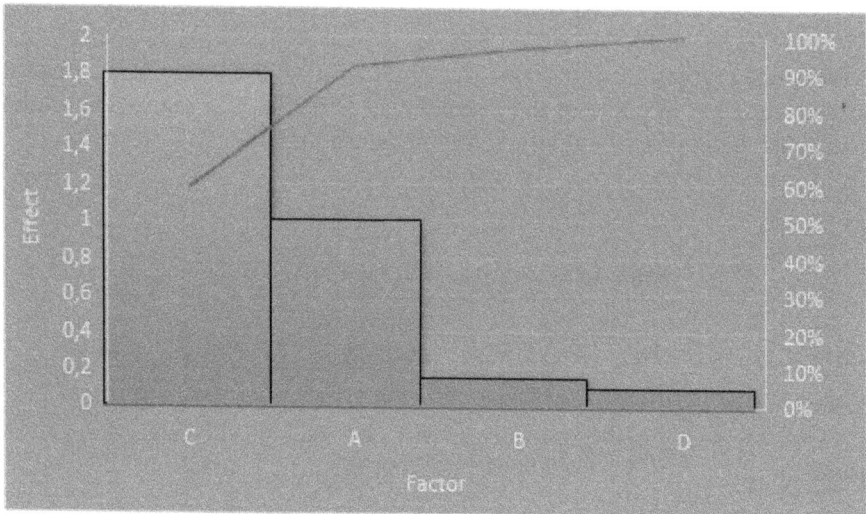

Fig. 4.18: Pareto plot

Main Effect

A main effect is the effect of a single independent variable on a dependent variable, ignoring all other independent variables. Based on the main effects plots in Fig. 4.19, the main effects are:

Main Effect = Mean Response at high (+) factor level – Mean Response at low (-) factor level

Factor A   Main Effect = 95 – 85 = **10**

130

Factor B	Main Effect = 90 − 81 = **9**
Factor C	Main Effect =100 − 80 = **20**
Factor D	Main Effect = 0 − 0 = **0**

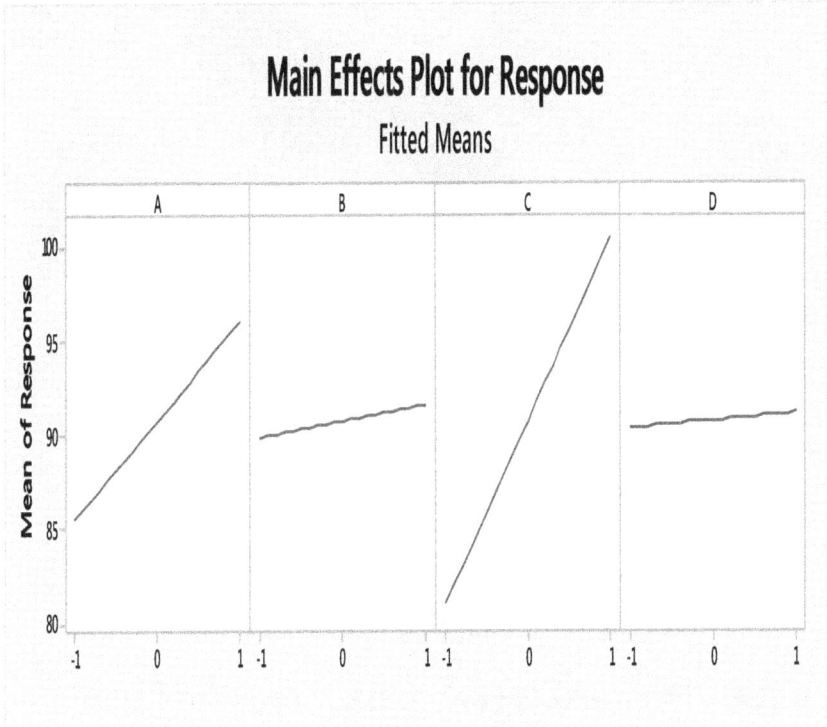

Fig. 4.19: Pareto plot

Response Surface Plot

The response surface plot (Fig. 4.20) shows that as the level of Factor A changes from low (-1) to high (+1), there is a corresponding increase in the response variable. However, the increase in response variable is steeper at high (+) level of Factor B than the low (-1) level. This therefore suggests that the effect of Factor A on the response variable is dependent on the level of Factor B, an indication of the presence of interaction between these two factors.

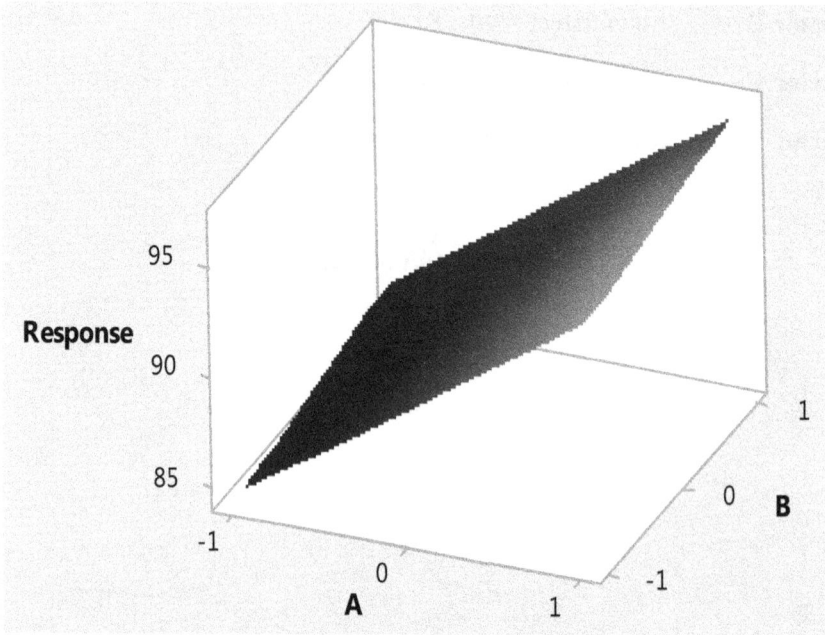

Fig. 4.20: Response Surface plot

Invention Inspirational Story

Charles Richard Drew- Father of Blood Banks

Charles Richard Drew was born on June 3, 1904, in Washington, D.C. He was a physician who developed ways to process and store blood plasma in blood banks. He directed the blood plasma programs of the United States and Great Britain in World War II.

Chapter 5: Response Surface Design

This chapter introduces the concept of response surface modelling (RSM), steps involved in RSM problem formulation and examples of response surface designs.

5.1. Introduction to Response Surface Models (RSM)

Response surface designs are primarily used for the estimation of interaction and quadratic effects of predictor variables in an experimental process. RSMs have the ability to provide information on the curvature of the response surface being investigated.

Applications of response surface models include identification of optimal process conditions, investigation of process problems and improvement of product or process robustness against uncontrollable external variables.

5.2. Examples of Response surface designs

5.2.1. Central Composite designs

Central composite designs are ideal for the development of second-order models. These models have the ability to estimate response surface curvature. These types of experimental designs are frequently used with response models of the second order. The design is made up of three types of points, namely axial, factorial and centre-points.

The three types of central composite designs are the central composite circumscribed (CCC), central composite face-centered (CCF) and the central composite inscribed (CCI) designs. Two of these designs CCI and CCC are rotatable. This means that if the design points are rotated, about the center-point the moments of the distribution of the design remain unchanged. This therefore makes CCC and CCI well suited for estimating coefficients in a second order model.

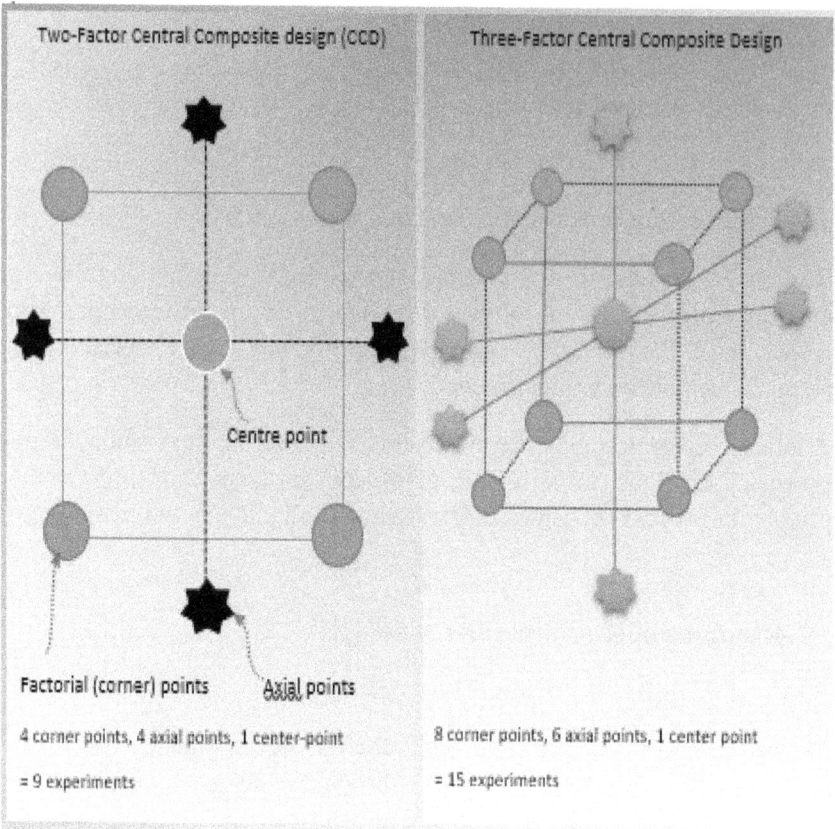

Fig. 5.1: Representation of 2-factor and 3-factor Central Composite Designs

a) *Central Composite Circumscribed (CCC)*

The axial points provide new extremes for the low and high factors conditions for all experimental factors. As a result of the alignment of the axial points, CCC designs have spherical symmetry. The factorial and axial points are located on the circumference of the circle. The accuracy of the design is therefore good over the entire design space. The distance of axial points (α) from the center-points depends on the number of factors in the full design as shown below;

$$\alpha = 2^{k/4}$$

$$k = number\ of\ factors$$

Distance of axial points (α) for two Central Composite Circumscribed Designs

(i)   2-factor   Central   Composite   Circumscribed   Design, $= 2^{2/4}$ $\alpha = 2^{1/2} = 1.414$

(ii)   3-factor   Central   Composite   Circumscribed   Design, $= 2^{3/4}$ $\alpha = 1.6818$

In general, Central Composite Circumscribed designs require 5 levels for each factor.

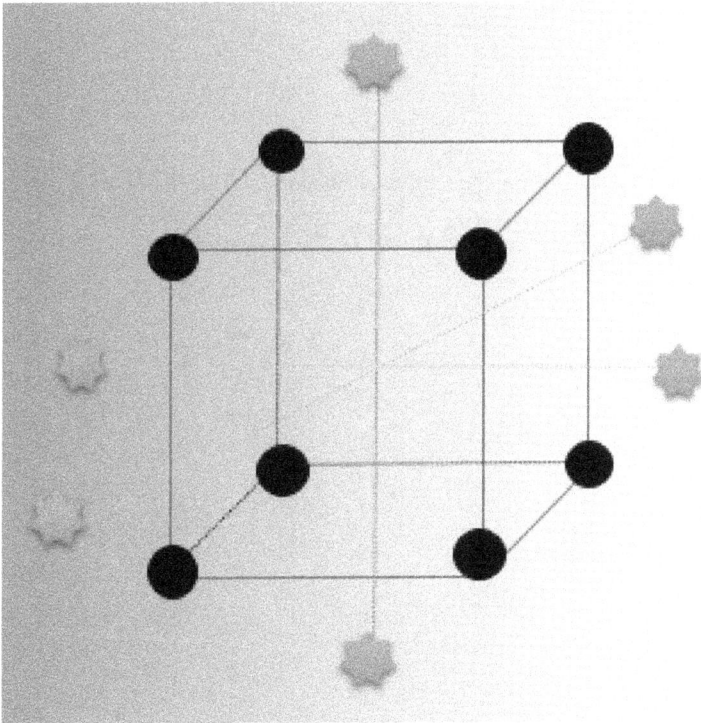

Fig. 5.2: Three factor Central Composite Circumscribed (CCC) designs with 2 levels

b) Central Composite Face-centered (CCF) design

The axial points in this design are located at the center of face of the factorial design. The CCF design requires only 3 levels for each

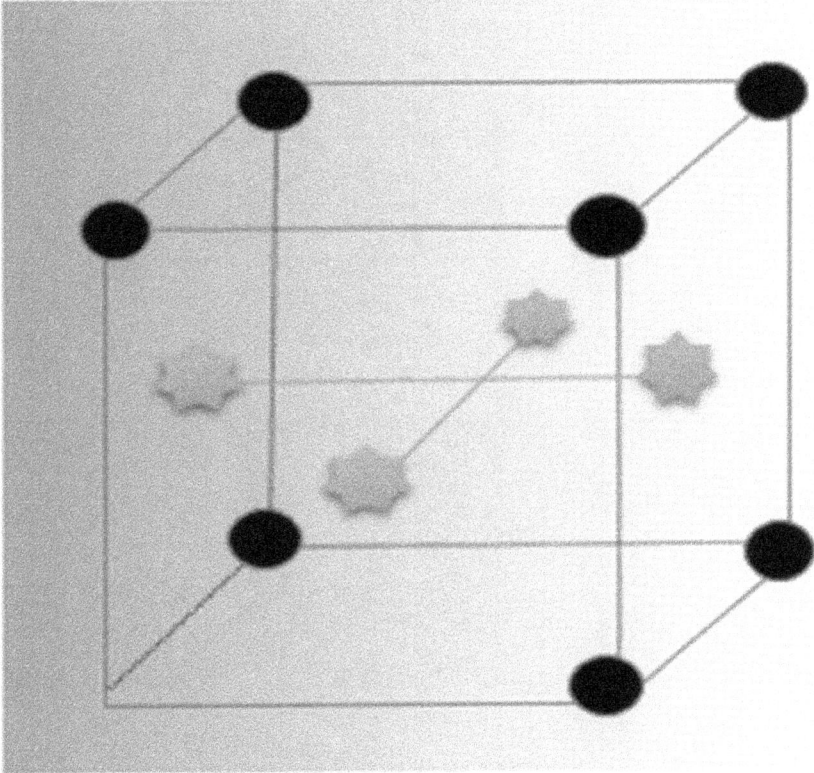

factor. The accuracy of the design is fair over the entire design space, but poor for pure quadratic coefficient.

Fig. 5.3: Three factor Central Composite Face-centered (CCF) design with 2 levels

c) *Central Composite Inscribed (CCI) design*

The central composite inscribed design has the distinguishing feature of using the factor settings as the axial points. The design requires 5 levels for each factor. The α value for this design is 1 and the design is rotatable as a result of every design point being equidistant from the center. The accuracy of the design is good over central subset of design space.

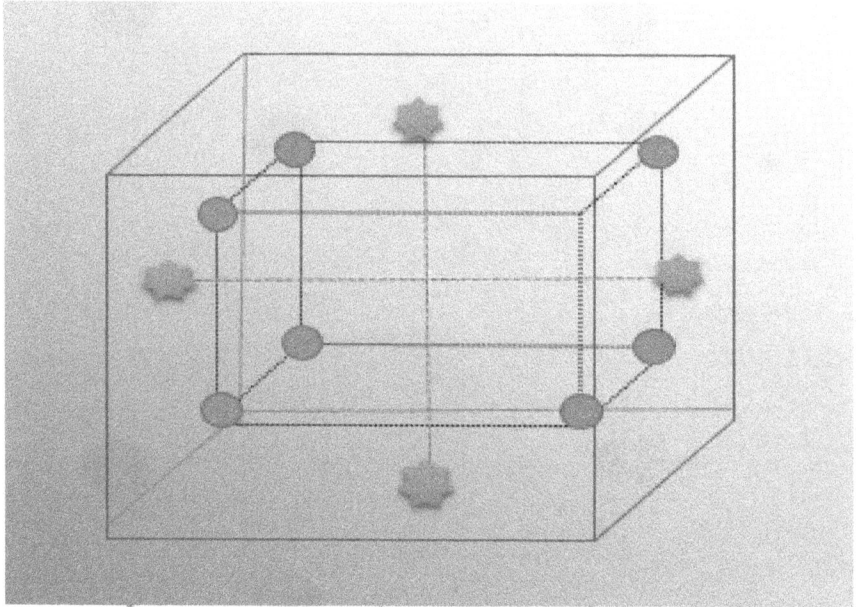

Fig. 5.4: Three Factor Central Composite Inscribed (CCI) design with 2 levels

Example 1: Central Composite Design Face-centered design

A central composite face-centered design was employed to study the main and interaction effects of three variables on one response.

a) *Development of a Central Composite Face-centered design matrix*

Using a face-centered design, a total number of 20 experimental runs, comprised of 8 cube points, 6 center-points and 6 axial points. The design matrix shows all possible combinations of high and low levels of each input factor. The high and low levels of each factor were coded as +1 and -1 respectively.

Table 5.1: Face-centered design matrix

Experiment	A	B	C	Response
1	1	1	1	83
2	-1	1	1	75
3	0	-1	0	9
4	0	0	0	9
5	-1	-1	1	25
6	1	0	0	9
7	1	-1	1	9
8	-1	0	0	28
9	1	1	-1	9
10	1	-1	-1	16
11	-1	-1	-1	50
12	0	0	0	26
13	0	1	0	34
14	0	0	1	9
15	0	0	0	9
16	0	0	0	16
17	-1	1	-1	29
18	0	0	0	13
19	0	0	-1	46
20	0	0	0	25

Coefficients Analysis

Based on the coefficient values, the null hypothesis that B and B*C are equal to zero is rejected. In other words B and B*C terms play a significant role in the regression model. However, all the other factors and their interactions do not play a significant role in the regression model, hence p values>0.05.

Table 5.2: Coefficient analysis

Term	Coefficient	Standard Error Of Coefficient	T-Value	P-Value
Constant	16.02	4.48	3.57	0.005
A	-8.10	4.12	-1.96	0.078
B	12.10	4.12	2.94	**0.015**
C	5.10	4.12	1.24	0.244
A*A	2.95	7.86	0.38	0.715
B*B	5.95	7.86	0.76	0.466
C*C	11.95	7.86	1.52	0.159
A*B	4.75	4.61	1.03	0.327
A*C	5.75	4.61	1.25	0.241
B*C	19.00	4.61	4.12	**0.002**

Model Summary

Standard deviation of error terms = 13.0368 $R^2 = 81.31\%$ $R^2adj = 64.49\%$

The summary statistics of the model indicate that the standard deviation of error terms is equal to 13.0368. $R^2 = 81.31\%$, meaning that an observed variation of the response, 81.31% of it is due to the model and only 18.69% is due to unexplained error.

Regression Equation

The regression line intersects the y-axis at 16.02. In other words, the response is 16.02 when all the factors are zero. The size of the coefficient indicates the magnitude of its effect on the response, all factors held constant. The sign of the coefficient shows the direction in which the coefficient influences the response. However, the size of the coefficient does not indicate whether the coefficient is statistically significant. Therefore p-values of the coefficients should be considered. Some coefficients in the model may be large, but not statistically significant. In this example, only one main factor B and one two-factor interaction (B*C) were found to be statistically significant (P<0.05).

Response = 16.02 - 8.10 A + 12.10 B + 5.10 C + 2.95 A*A + 5.95 B*B + 11.95 C*C + 4.75 A*B+ 5.75 A*C + 19.00 B*C

Pareto Chart

The purpose of the Pareto chart is to highlight the most important among the three factors on the response. Therefore, the Pareto chart assists in visualizing the factors that comprise the 20% but explain 80% of the changes observed in the response. One main factor B and one two-factor interaction term B*C were found to be responsible for 80% of the changes observed in the response.

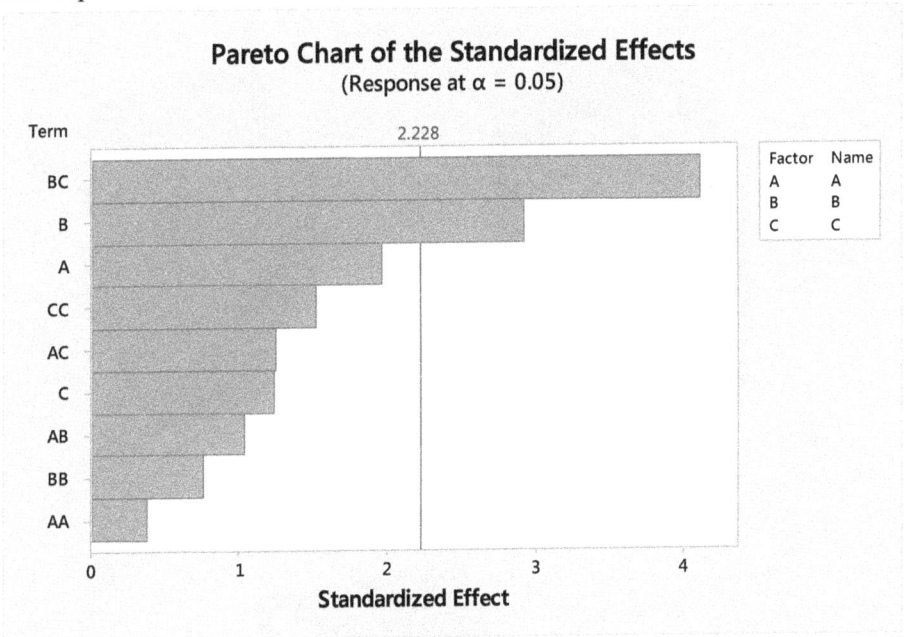

Pareto Chart of the Standardized Effects
(Response at α = 0.05)

Factor	Name
A	A
B	B
C	C

Fig. 5.5: Pareto chart of Response

Response Surface

141

Response surface methods are designs are ideal when working with continuous variables to find an optimum or to describe the response. In other words, response surface designs are optimization models. In this example, three predictor variables are varied in accordance with a central composite face-centered design (CCF) with a total of 20 experimental runs. CCF designs support a quadratic model. Large squared terms of A, B and C are also observed and the response surface plot of the response is curved. This curvature displayed a falling ridge response surface (Fig.5.6).

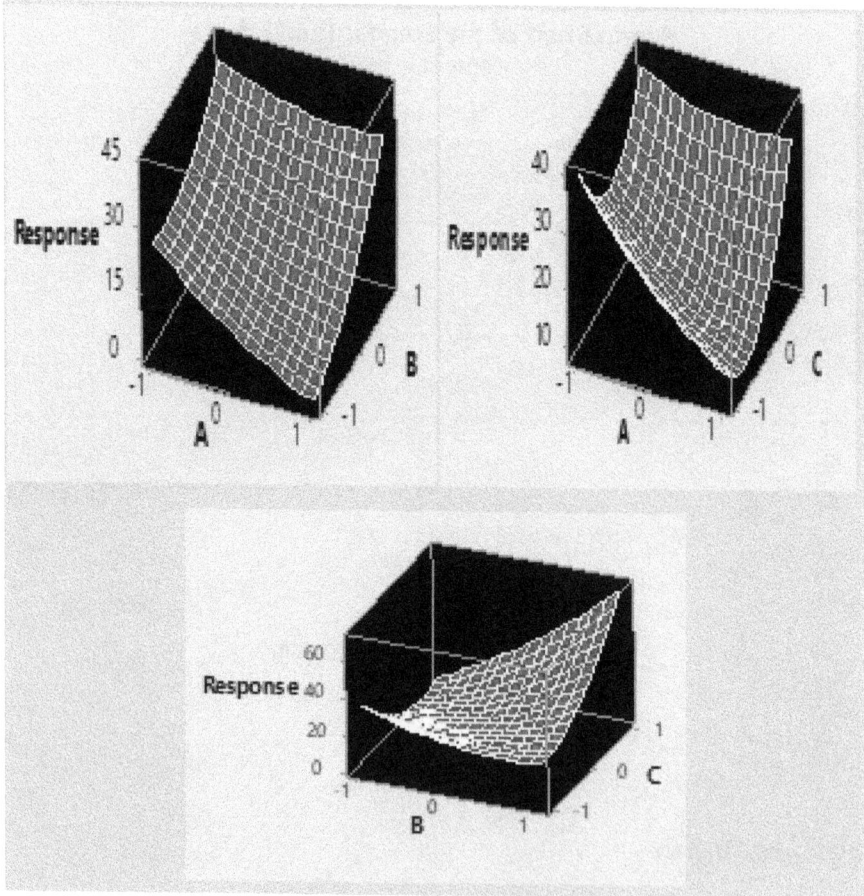

Fig. 5.6: Response surface plots

Main Effects Plots

Main effects plots confirm the model curvature. These plots suggest that a change in factor level of A from low (-1) to high (1) results in a gradual decrease in the response. However, a change in factor B from low (-1) to high (1) results in an increase in response. In the case of factor C, the curvature is more pronounced. A change in factor C from low (-1) to mid-point (0) results in a decrease in the response. However, as factor C increases past mid-point (0) towards its higher level (1), an increase in the response is observed. Therefore, responses that exhibit the behaviour in Fig. 5.7 require a minimum of three levels for each factor to fully characterize the behaviour of the response variable. In general, an addition of a center point to a two-level design should be sufficient.

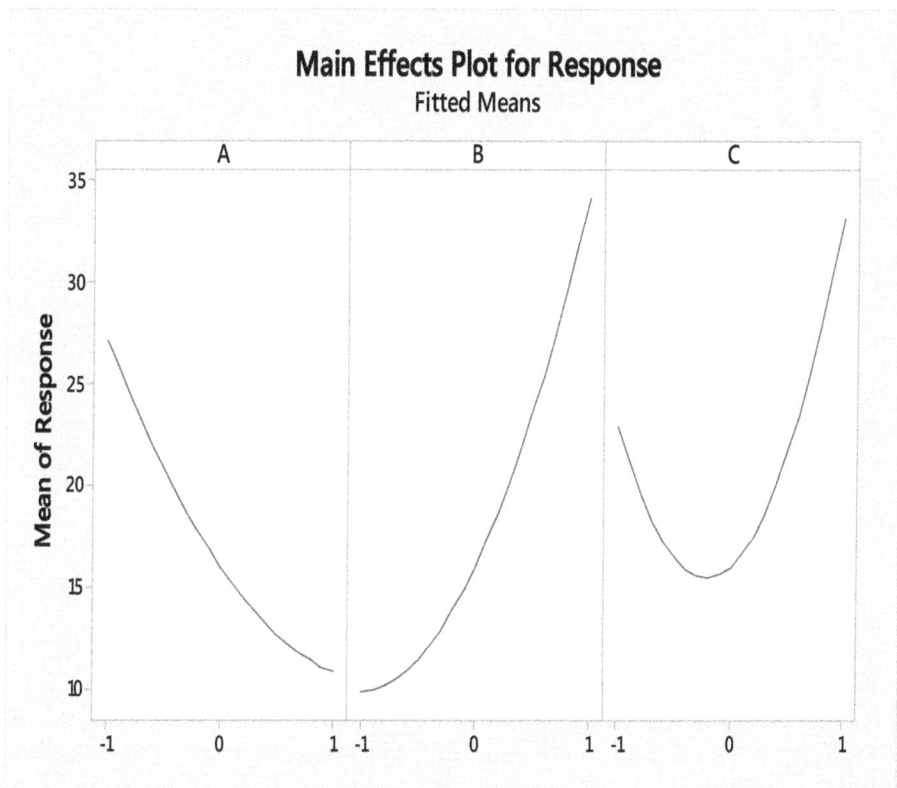

Fig. 5.7: Main effects plots

Interaction Plots

The plots indicate that the effects of these factors on the response are dependent on the level of the other factors, hence confirming the interaction between factors. The strongest interaction was between factors B and C.

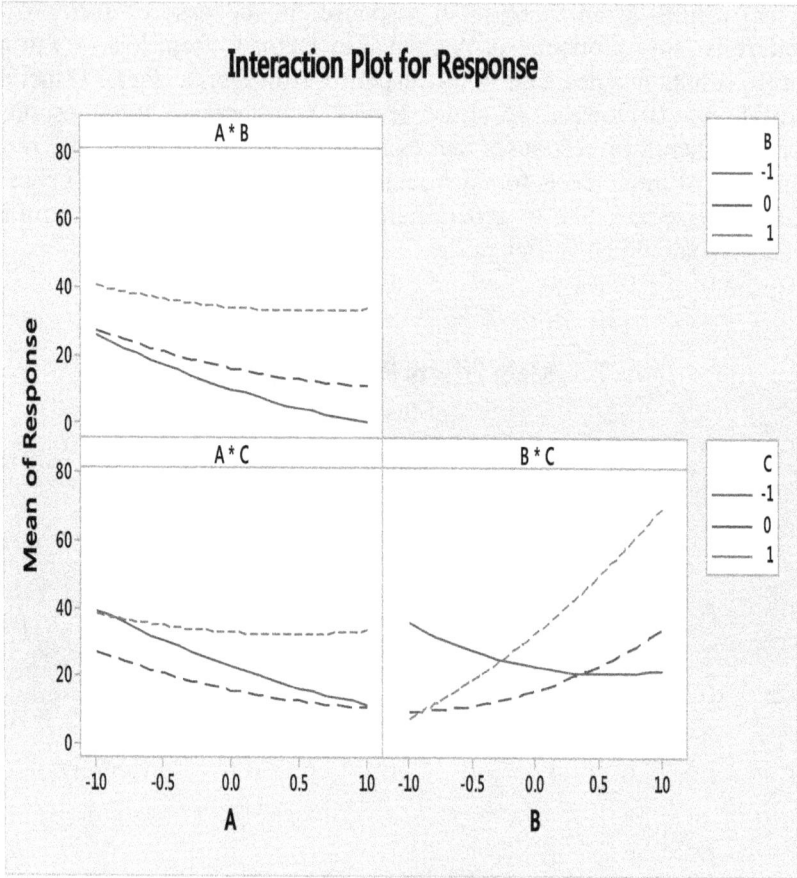

Fig. 5.8: Interaction plot of factors

Contour

The contour plots present a rising ridge surface. As the colour gets darker, the response increases. This means the response increases as factors A, B and C are varied from low (-1) to high (1). These contour plots are based on a regression model.

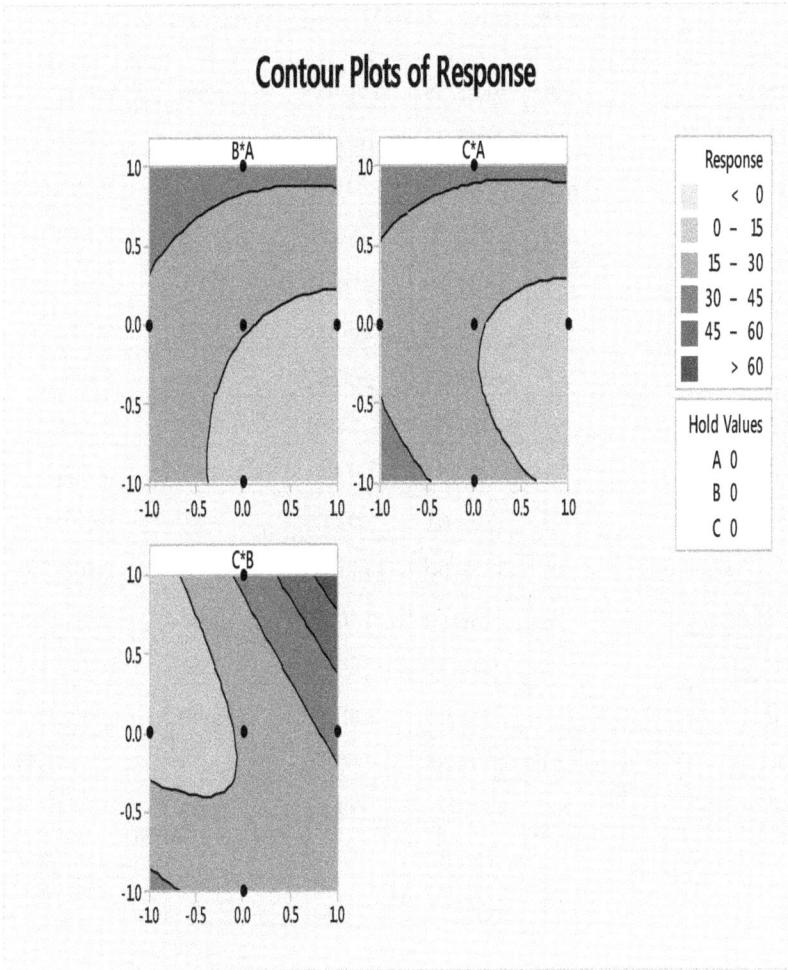

Fig. 5.9: A Response Surface with a falling ridge

Analysis of Variance

The variation on the response due to the model is statistically significant (P<0.05), while the variation due to residuals is not significant (P>0.05).

Table 5.3: Analysis of Variance (ANOVA)

Source	DF	Adj SS	Adj MS	F-Value	P-Value
Model	9	7393.37	821.49	4.83	0.011
Linear	3	2380.30	793.43	4.67	0.027
A	1	656.10	656.10	3.86	0.078
B	1	1464.10	1464.10	8.61	0.015
C	1	260.10	260.10	1.53	0.244
Square	3	1680.07	560.02	3.30	0.066
A*A	1	24.01	24.01	0.14	0.715
B*B	1	97.51	97.51	0.57	0.466
C*C	1	393.01	393.01	2.31	0.159
2-Way Interaction	3	3333.00	1111.00	6.54	0.010
A*B	1	180.50	180.50	1.06	0.327
A*C	1	264.50	264.50	1.56	0.241
B*C	1	2888.00	2888.00	16.99	0.002
Error	10	1699.58	169.96		
Lack-of-Fit	5	1412.25	282.45	4.92	0.053
Pure Error	5	287.33	57.47		
Total	19	9092.95			

Validity of Error Assumptions (Fig. 5.10)

a) Normal Probability Plot

This plot attempts to confirm the assumption of normality of errors. Most of the points in the plot are distributed around the diagonal line, suggesting the error terms are approximately normal. This confirms that the assumption of normality is valid.

146

b) Residuals vs Fitted values

This is a plot of the error terms against the fitted values. This plot tests the assumption that error terms have a mean equal to 0. The residuals versus fits plot is used verify the assumption that the residuals have a constant variance i.e. that the residuals are spread randomly around the 0 line, suggesting that relationship is linear. Residuals display a horizontal band around the 0 line, suggesting that the variances of error terms are equal. The plot demonstrates that almost half of the points lie above and below the zero line, hence confirming the assumption that the error terms have a mean of zero.

c) Histogram of Residuals

The histogram of residuals is used determine whether the data are skewed or whether outliers exist in the data. A symmetric bell-shaped histogram evenly distributed around zero shows that the normality assumption is true. In this example, the histogram of residuals confirms the normality of errors assumption.

d) Residuals vs Observation order

The residuals bounce randomly around the residual = 0 line. This indicates that the residuals are demonstrating normal random noise around the residual = 0 line, implying that there is no serial correlation.

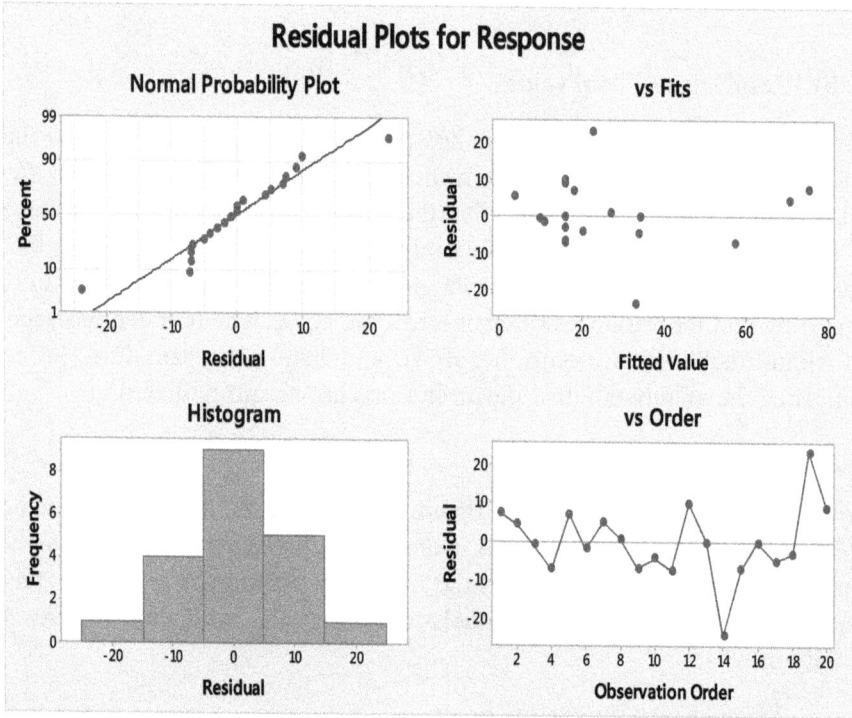

Fig. 5.10: Test for validity of assumptions

Problem 2: The research and development department of a mouthwash manufacturing company assume that the optimal conditions for the manufacture of the mouthwash are:

a) Abrasive (30-50%)
b) Humectants (20-40%)
c) Thickening agent (20-30%)

Currently, the company is manufacturing the mouthwash using 40% abrasive, 30% humectant and 25% thickening agent. The product developers decide to explore the current conditions in the range [Low=30, High=50] for abrasive (X_1), [Low=20, High=40] for humectants (X_2) and [Low=20, High=30] for thickening agent (X_3) to determine how the changes in abrasive, humectant and thickening agent affect the consistency of the mouthwash. The consistency of the mouthwash was measured on

148

a scale of 0-100. A Box-Behnken design was selected for the development of the design matrix, based on its economical number of experimental runs.

Development of a Box-Behnken Design matrix

The Box-Behnken designs are quadratic designs that do not contain an embedded factorial or fractional factorial design. In Box-Behnken designs, the experimental treatments are at the mid-points of the edges of the designs space. Box-Behnken designs require 3 levels for each factor.

Table 5.4: Box-Behnken Design matrix

Abrasive	Humectant	Thickening Agent	Consistency
0	0	0	33.95
1	-1	0	36.35
0	1	1	35.10
0	0	0	37.20
1	0	1	35.30
0	-1	-1	35.60
-1	0	-1	36.15
-1	-1	0	35.45
-1	1	0	35.90
1	1	0	32.00
0	1	-1	31.00
0	0	0	28.00
1	0	-1	27.00
0	-1	1	25.00
-1	0	1	30.00

Regression Analysis

Regression analysis aims to develop a model that describes the statistical relationship between one or more predictor variables and the response variable.

The p-value for each term in the regression expression tests the null hypothesis that the coefficient is equal to zero. P-values< 0.05 result in the rejection of the null hypothesis, suggesting that the predictor variable has a significant effect on the regression model.

However, P values>0.05, mean that the predictor variable is not responsible for the variations in the response variable. In Table 5.5, none of the factors and their interactions have a statistically significant influence on the response variable. As shown in Fig. 5.11, the error bars are greater than the effects of the factors.

Table 5.5: Coefficient analysis

Term	Coefficient	Standard Error Of Coefficient	T-Value	P-Value
Constant	33.1	2.1	16.0	0.000
Abrasive	-0.9	1.3	-0.7	0.529
Humectant	0.2	1.3	0.2	0.881
Thickening Agent	-0.5	1.3	-0.4	0.685
Abrasive*Abrasive	1.2	1.9	0.6	0.562
Humectant*Humectant	0.7	1.8	0.4	0.715
Thickening Agent*Thickening Agent	-2.09	1.86	-1.1	0.312
Abrasive*Humectant	-1.20	1.79	-0.7	0.532
Abrasive*Thickening Agent	3.61	1.79	2.02	0.099
Humectant*Thickening Agent	3.68	1.79	2.05	0.095

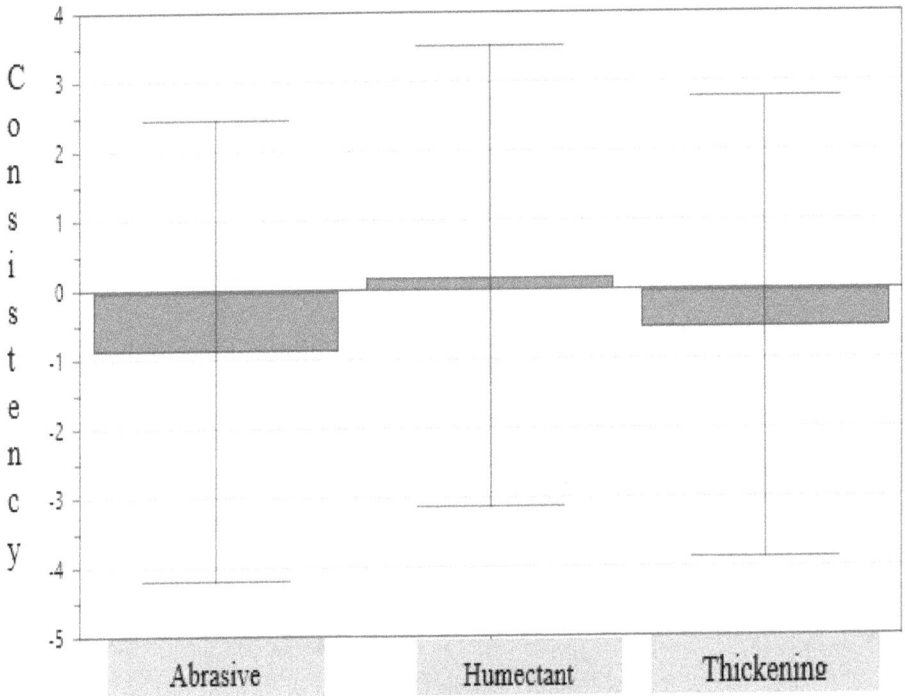

Coefficient Plot for Consistency

Fig. 5.11: Coefficient plot

Model Summary

Standard deviation of error terms = 3.57836 $R^2 = 69.44\%$ $R^2adj = 14.43\%$

The summary statistics of the model indicate that the standard deviation of error terms is equal to 3.57836. $R^2 = 69.44\%$. This means that an observed variation of the response, 69.44% of it is due to the model and only 30.56% is due to unexplained error.

Regression Equation

Consistency	=	33.05 - 0.86 Abrasive
		+ 0.20 Humectant
		- 0.54 Thickening Agent
		+ 1.16 Abrasive*Abrasive
		+ 0.72 Humectant*Humectant
		- 2.09 Thickening Agent*Thickening Agent
		- 1.20 Abrasive*Humectant
		+ 3.61 Abrasive*Thickening Agent
		+ 3.68 Humectant*Thickening Agent

Pareto Chart

Pareto analysis helps to identify the significant few factors that affect the response variable. In other words, the Pareto chart analyses the data to rank the effects of the factors on the response from the most to least. Based on Fig. 5.12, the most influential single factor on consistency is Abrasive (Abras), followed by Thickening (Thick) and lastly Humectant (Hum). However, the most significant effects on consistency are the two-factor interactions Hum*Thick, followed by Abras*Thick and Thick*Thick.

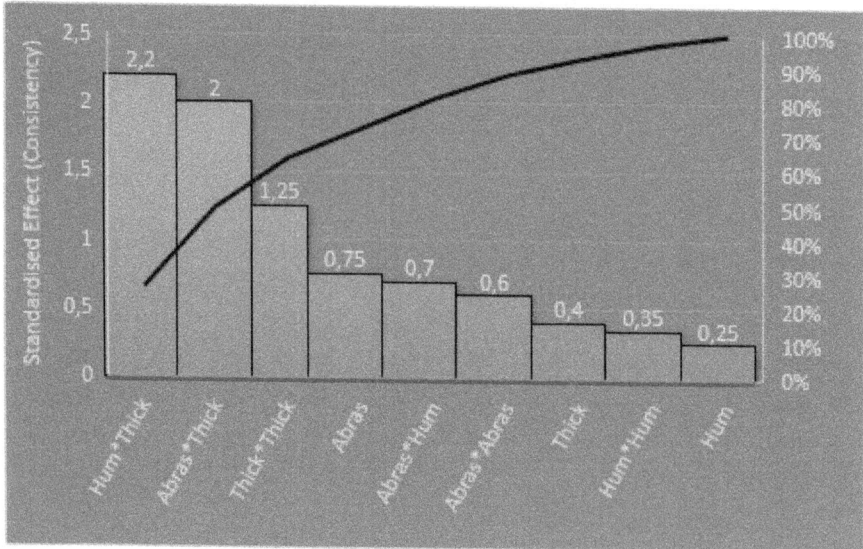

Fig. 5.12: Pareto chart of the standardized effects

152

Response Surface Plot

The response surface indicate the changes in the response variable (consistency), with changes in the predictor variables. Response surface plots indicate the curvature caused by quadratic terms in the model (Fig. 5.13).

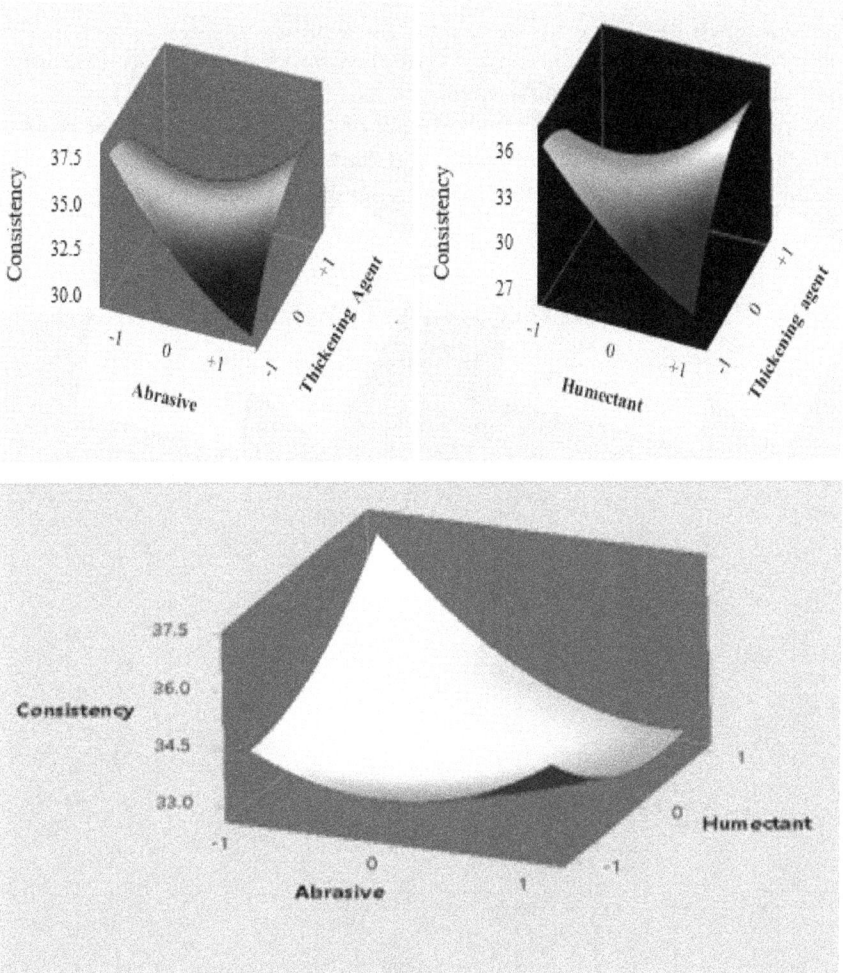

Fig. 5.13: Response surface plots showing changes in the consistency (response) and the predictor variables (humectant, abrasive and thickening agents) are varied.

Main Effects Plots

Main effects plots confirm the model curvature seen in the response surface plots. From Fig.5.14, it is clear that as the predictor variable (Abrasive) changes from its low level (-1) to the mod-point level (0), the consistency of the response changes from 35 to 33. However, as the level of the abrasive changed from 0 to higher level (1), the consistency increased to about 33.5. A similar pattern is observed with humectant. However, the change of the thickening agent from its low level (-1) to its middle value (0), an increase in consistency of the response is observed from 30 to 33. After the middle point (0), the change of the thickening agent towards the higher level (1) results in a corresponding decrease in the consistency of the response. In other words, the main effects plots of the abrasive and humectant demonstrate a falling ridge curve compared to the rising ridge shown by the thickening agent.

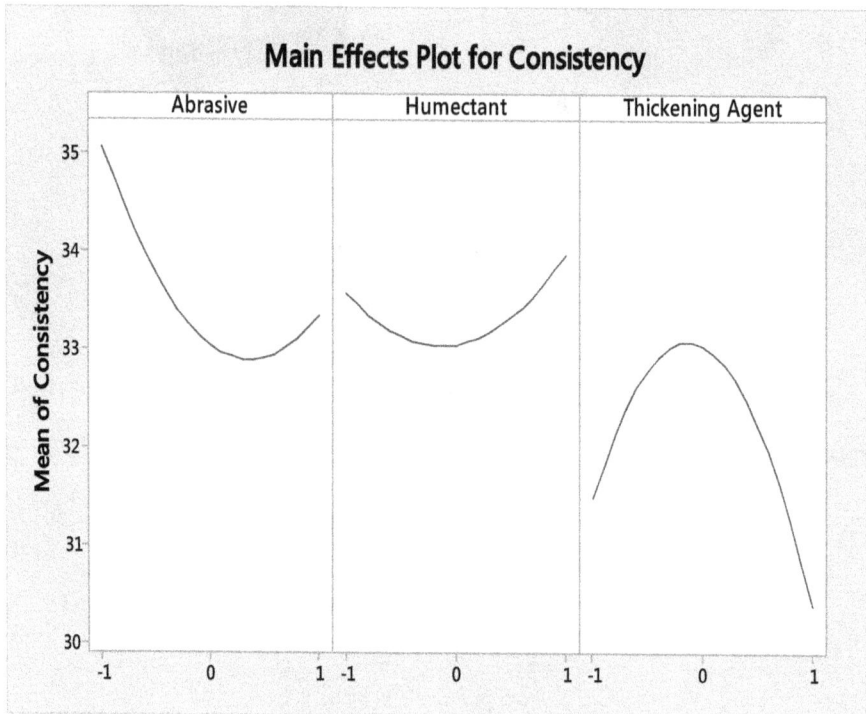

Fig. 5.14: Main effects plots

Interaction Plots

The interaction plots indicate that the effects of the predictor variables on the response are dependent on the level of the other factors, hence confirming the existence of interaction between predictor variables. Parallel lines indicate that there is no interaction between the predictor variables, while non-parallel lines demonstrate the presence of interaction between factors. The less parallel the lines are, the greater the degree of interaction.

Fig. 5.15, illustrates non-linear cases with interaction. Greatest interaction is observed for abrasive*thickening agent and humectant*thickening agent. Comparatively weaker interaction is observed between abrasive*humectant.

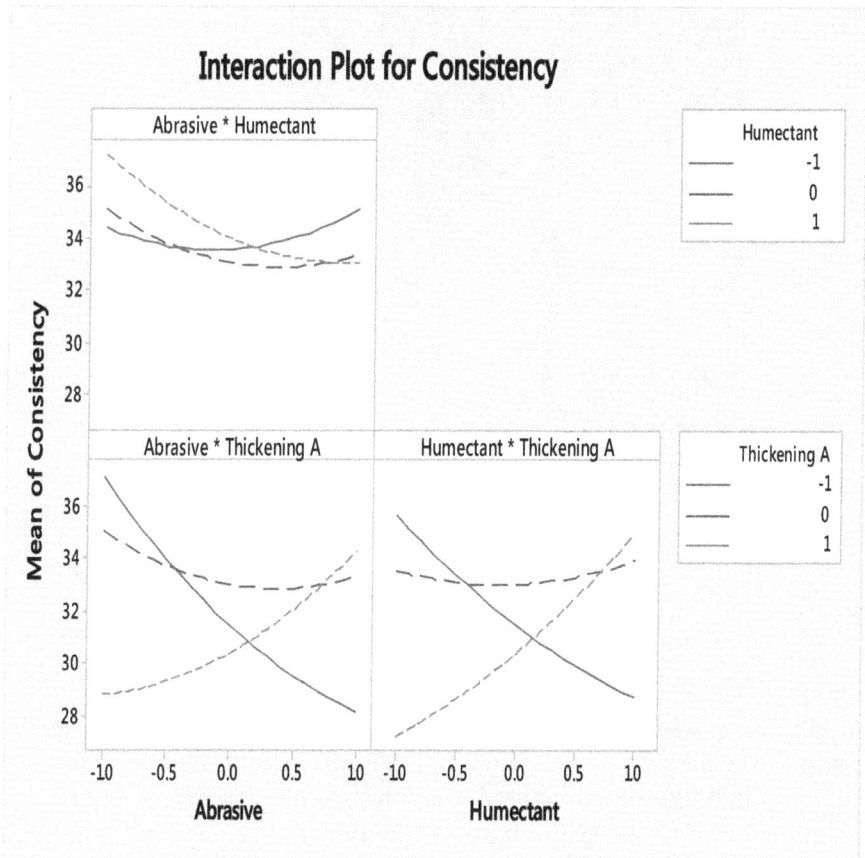

Fig. 5.15: Non-linear cases with interactions

Contour

The contour plots confirm the findings from the response surface plots. Two distinct contour topologies are evident namely "Falling" Ridge and "Saddle" for humectant*abrasive and thickening agent*abrasive respectively.

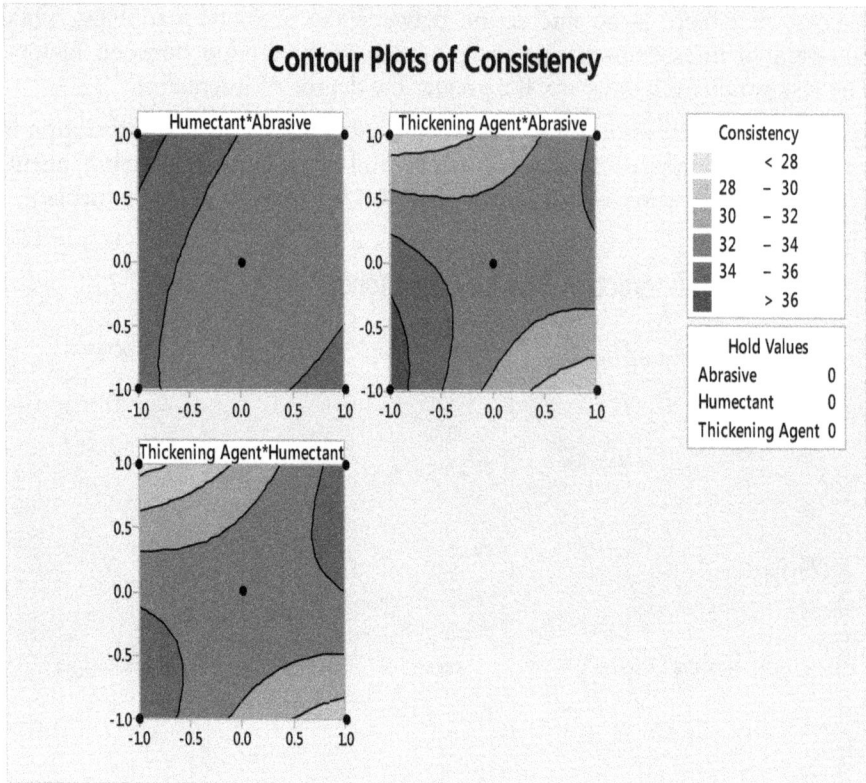

Fig. 5.16: Contour plots

Analysis of Variance

To determine whether the regression model explains the variation in the response, the p-value for the model is compared to the significance level of 0.05. The null hypothesis is that the regression model does not explain the changes in the response. In this study, the p-value for the model is >0.05, suggesting that the regression model does not explain the variation in the response. Each of the terms' coefficients is not statistically different from zero, with p>0.05. This means that none of the individual main effects and their interactions have a statistically significant effect on the changes in the response variable. The lack-of-fit error was also not statistically significant.

Table 5.6: Analysis of Variance (ANOVA)

Source	DF	Adj SS	Adj MS	F value	P-Value
MODEL	9	145.5	16.2	1.3	0.419
Linear	3	8.6	2.9	0.2	0.877
a) Abrasive	1	5.9	5.9	0.5	0.529
b) Humectant	1	0.3	0.3	0.02	0.881
c) Thickening Agent	1	2.4	2.4	0.20	0.685
Square	3	24.9	8.3	0.65	0.617
a) Abrasive*Abrasive	1	4.9	4.9	0.39	0.562
b) Humectant*Humectant	1	1.9	1.9	0.15	0.715
c) Thickening Agent*Thickening Agent	1	16.2	16.2	1.26	0.312
2-Way Interaction	3	112	37.3	2.92	0.140
a) Abrasive*Humectant	1	5.8	5.8	0.45	0.532
b) Abrasive*Thickening Agent	1	52.2	52.2	4.08	0.099
c) Humectant*Thickening Agent	1	54.0	54.0	4.22	0.095
ERROR	5	64.0	12.8		
Lack-of-Fit	3	20.5	6.8	0.31	0.819
Pure Error	2	43.535	21.8		
Total	14	209.493			

Validity of Error Assumptions

The test for validity of assumptions might assist us determine the reason for failure of the regression model to fit the experimental data.

a)   Normal Probability Plot

This plot attempts to test the assumption of normality of errors. The points are not precisely distributed around the diagonal line, suggesting that the error terms are not approximately normal. It can be concluded that the assumption of normality is not necessarily valid.

b)   Residuals vs Fitted values

This is a plot of the error terms against the fitted values. This plot tests the assumption that error terms have a mean equal to 0. The plot demonstrates that more points lie above than below the zero line, hence failing to confirm the assumption that the error terms have a mean of zero.

c)   Histogram of Residuals

Histogram of residuals are used to determine whether the data are normally distributed, skewed or whether there are outliers in the data.

d)   Residuals vs Observation order

In this residuals vs order plot, residuals do not appear to be randomly spread around the residual = 0 line. Residuals appear to systematically decrease as the observation order increase. This suggests that error terms are dependent on time.

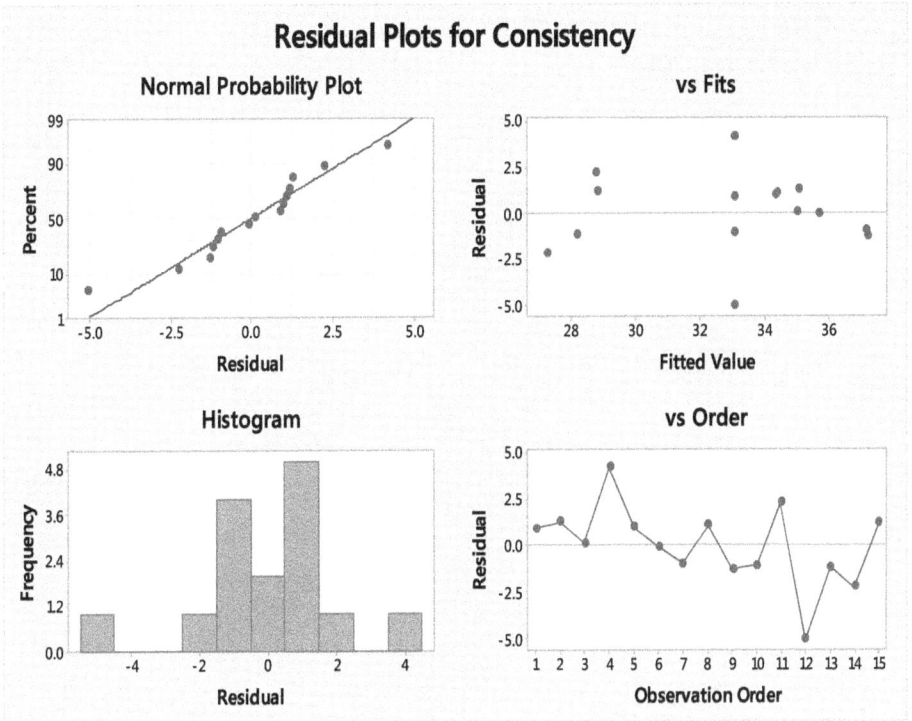

Fig. 5.16: Checking the validity of assumptions

Invention Inspiration Story

Patricia Bath

An inventor and opthamologist from Harlem, New York. In 1986, Bath invented the Laserphaco Probe, improving treatment for cataract patients. She patented the device in 1988. In 1976, Bath co-founded the American Institute for the Prevention of Blindness.

www.ingramcontent.com/pod-product-compliance
Lightning Source LLC
Chambersburg PA
CBHW022040190326
41520CB00008B/664